T0336095

Synbiotic Foods

Developed using different matrices such as dairy, cereals, legumes, fruits, and vegetables, synbiotic food products combine the benefits of both probiotics and prebiotics. This book is a clear and comprehensive guide to the core concepts of synbiotic foods and associated technological advancements and applications across food groups. Using clear, scientific language, this book equips students with in–depth knowledge of synbiotic products, processing methods, applications, and acceptance.

This is a pioneering textbook on synbiotic foods, being the first of its kind to include the following features:

- Explores fundamental aspects across various matrices
- Chapter summaries using ten concise bullet points
- Multiple choice questions (MCQs) for aiding study for national-level competitive exams
- Short and long descriptive answer type questions for comprehensive exam preparation

Designed as a one-stop resource, this book particularly appeals to undergraduate and postgraduate students of Food Science, Food Technology, Food Biotechnology, and Food Microbiology.

FUNCTIONAL FOODS AND NUTRACEUTICALS SERIES

Series Editor:
John Shi, Ph.D.
Guelph Food Research Center, Canada

For more information about this series, visit https://www.routledge.com/
Functional-Foods-and-Nutraceuticals/book-series/CRCFUNFOONUT

Synbiotic Foods
Significance, Applications, and Acceptance

Smriti Chaturvedi and Snehasis Chakraborty

CRC Press
Taylor & Francis Group
Boca Raton London New York

CRC Press is an imprint of the
Taylor & Francis Group, an **informa** business

First edition published 2024
by CRC Press
2385 NW Executive Center Drive, Suite 320, Boca Raton FL 33431

and by CRC Press
4 Park Square, Milton Park, Abingdon, Oxon, OX14 4RN

CRC Press is an imprint of Taylor & Francis Group, LLC

© 2024 Smriti Chaturvedi and Snehasis Chakraborty

ISBN: 978-1-032-29337-0 (hbk)
ISBN: 978-1-032-30242-3 (pbk)
ISBN: 978-1-003-30410-4 (ebk)

DOI: 10.1201/9781003304104

Typeset in Caslon
by MPS Limited, Dehradun

Contents

Preface

Synbiotic Foods: Significance, Applications, and Acceptance covers a significant portion of a novel concept in food processing and technology, mainly in the form of an elective for both bachelor's and master's students in Food Science, Food Technology, Food Biotechnology, or Food Microbiology. This textbook explores the fundamental aspects of various synbiotic foods in the domain of different food groups to facilitate classroom teaching as well as learning processes.

The book's 11 chapters, written in scientific but straightforward language, encompass all the synbiotic products in depth, from basic concepts to technological advances. The intent of this book is to provide a complete study material or single information source for the students while covering the fundamental aspects of these synbiotic foods, their processing, application, and acceptance coupled with questions on the same.

The book begins with a general introduction dealing with the definitions of probiotics, prebiotics, and then the concept of synbiotics. The fundamental mechanism of all of these inside the human gut has also been precisely briefed. Concomitantly, various matrixes that serve as potential vehicles for synbiotic foods are summarized in individual chapters. These matrices are dairy, fruits, vegetables, cereals, and legumes. For every chapter (2–6), the general structure is the same, which includes processing, associated health benefits, and different products under each matrix. It is followed by a bullet-point summary of the entire chapter. In the later chapters (7–10), the potential of synbiotic foods in powder form, the screening and evaluation of synbiotics, consumer perception, and commercial aspects, and other challenges associated with synbiotic food products are discussed. In Chapter 11, a comprehensive detail of different synbiotic products is provided in a tabular form. For each chapter, there are a set of multiple choice questions (MCQs) and short and long answer type descriptive questions.

We believe this book can be a ready reckoner for undergraduate and postgraduate students in Food Science and Technology, Food Biotechnology, or Food Microbiology to learn various synbiotic foods and their processing. We shall be happy to receive suggestions for improving the first edition of this textbook to incorporate into the next edition.

Smriti Chaturvedi and Snehasis Chakraborty

About the Authors

Dr. Smriti Chaturvedi is a post-doctoral candidate from the School of Engineering, University of Guelph, Canada. Her research focuses on synbiotic foods, probiotics and prebiotics, legume-based non-dairy food products, product optimization, and food safety.

Dr. Snehasis Chakraborty is an assistant professor of food technology at the Institute of Chemical Technology, Mumbai, India. He is a visiting scientist at Kansas State University, USA. His research area includes synbiotic foods, non-thermal and advanced thermal processing of foods, process optimization, kinetic modeling, shelf-life study, and sensory analysis. He is the academic editor of the *Journal of Food Processing and Preservation and the Journal of Food Biochemistry*. He is also an editorial board member of the *Applied Food Research*.

INTRODUCTION TO SYNBIOTICS

1.1 Probiotics

The word 'probiotic' is obtained from a Greek word that means 'for life.' In microbiology, probiotics are living microorganisms that are non-pathogenic in nature and exert beneficial effects on the host organism. As per the Food and Agriculture Organization (FAO) of the United Nations and the World Health Organization (WHO), probiotics are 'living microorganisms which, when administered in adequate amounts, confer health benefits to the host' (Gibson et al., 2017). Along with this, a true probiotic must also have the following characteristics:

1. Ability to tolerate and survive intestinal conditions (gastric pH, enzymes, bile salts, etc.)
2. Exhibit antagonism against pathogens
3. Stimulate the immune system
4. Exert beneficial effects on the host
5. Adhere to the epithelial cell lines
6. Possess antimicrobial activity for prevention against pathogenic microbes
7. Be stable and viable for reasonable durations under storage and upon processing treatments.

The most common probiotic microorganisms used in the commercial food and pharmaceutical industry belong to the *Lactobacillus* genus, such as *Lactobacillus acidophilus, L. casei, L. rhamnosus,* and *L. reuteri*; other genera include *Bacillus coagulans, Enterococcus faecium* SF68, *Escherichia coli* strain *Nissle* 1917, and yeast *Saccharomyces boulardii.*

To exhibit the required health benefits, any food product claimed as a 'probiotic' must have a minimum dose, i.e., 10^6 colony-forming units (CFU)/mL of the viable probiotic microorganism at the time of consumption. However, considering numerous factors like the amount ingested, processing conditions, and the effect of storage on the viability of probiotics, consumption of probiotics in a range of 10^8–10^9 CFU/g (as recommended by the U.S. Food and Drug Administration, FDA) is usually attained to deliver the required benefits to the host.

Probiotics are majorly known to modulate and improve gut functionality, though recent studies have also shown the positive effects of probiotics in enhancing immunity, refining brain functioning, reducing cholesterol levels, and helping metabolic homeostasis in the body. But in order to exert these benefits, the probiotic must reach the human gastrointestinal (GI) tract in an active form. For this purpose, the probiotics need a medium or carrier to reach the gut. In general, food products like beverages and yogurts and pharmaceutical products like capsules and pills are consumed to serve this need. However, there is a need to explore more dietary options for the intake of probiotics in easier ways. For this reason, the selection of an appropriate matrix for the delivery of probiotics is crucial. Moreover, the sensitivity of probiotics to higher temperatures and heat also creates another challenge in the selection of appropriate food products whose processing does not kill the viability of these beneficial microbes. Food products such as fruits and vegetable juices, milk, and dairy products like yogurt, ice cream, etc., are the most commonly available probiotic products. Other emerging food-based delivery matrixes for probiotics include cereal and legume-based beverages. The addition of probiotics in such food products allows their easy transfer, retaining their activity, and delivering the desirable beneficial health effects.

1.2 Prebiotics

Unlike probiotics, prebiotics are non-viable substances that provide nutrition to the indigenous gut microbiota as well as administered probiotics. As per the definition by the International Scientific Association for Probiotics and Prebiotics, 'prebiotic' is a substrate that is selectively utilized by host microorganisms, conferring a health benefit (Hill et al., 2014). A prebiotic must also possess the following properties:

1. Resistant to the acidic pH of the stomach, preventing it from getting absorbed or hydrolyzed by gastric enzymes in the GI tract

2. Easily fermented by gut microorganisms and controls the modulation in gut microflora
3. Active at lower dosages and
4. Stable at acceptable storage and processing conditions

Some of the most common sources of naturally occurring prebiotics include soybeans, artichokes, raw oats, chicory roots, yacon, onion, garlic, unrefined wheat, and barley. The prebiotic in these plant sources is present in the form of inulin, fructooligosaccharides (FOS), galactooligosaccharides (GOS), and raffinose family oligosaccharides (RFOs) (Adebola et al., 2014). These carbohydrates serve as a source of energy to gut microorganisms, which in turn confer several health benefits such as providing relief from inflammation associated with intestinal bowel disorder, reducing the occurrence of diarrhea, preventing obesity, lowering some risk factors for cardiovascular disease, preventing colon cancer, and enhancing the bioavailability and uptake of minerals.

Although we consume either the prebiotic source or probiotic food separately, their effectiveness is not seen in the body properly. Hence, the incorporation of both of these individual components at one place, in adequate quantities, to confer desirable benefits is required. For this purpose, the concept of 'synbiotic' in the food industry is novel and holds the potential to serve as an innovative dietary food option.

1.3 Synbiotics

The term 'synbiotic' defines a combination of probiotic(s) and prebiotic(s) that act synergistically and stimulate the growth and metabolism of intestinal microbes, thereby exerting beneficial effects on human health. While the prebiotic component should selectively allow the growth of probiotics by serving as a source of energy and food, the probiotic component must also exhibit the required health benefits in the presence of the prebiotics. Therefore, selecting a proper combination of both components in one product is essential to ensure a higher combined effect. The reason behind the development of a synbiotic product is to overcome the possible survival difficulties faced by probiotics as soon as they reach the human GI tract (low pH because of gastric acid and bile salts). Hence, a synbiotic product aims to positively affect the host by increasing the survival rate of probiotics and indigenous microbiota in the gut and selectively encouraging the growth and metabolism of probiotics in the GI tract.

1.4 Mechanism

1.4.1 Probiotics

Probiotics are known to positively impact human gut health by exhibiting several regulatory functions. They regulate the gut barrier integrity, sustain the intestinal microbiota homeostasis, and control immune system-based responses (Figure 1.1). Probiotics prevent the host from pathogenic invasion and play a crucial role in decreasing the risks of diabetes, inflammatory bowel disease (IBD), obesity, liver and cardiovascular cancer, and disorders associated with the central nervous system. These beneficial microorganisms can exhibit these health benefits primarily because of the major mechanisms followed by them.

The mechanisms or the modes of action followed by probiotics can be of two types:

1. Direct antagonistic inhibition between the commensals and pathogens in the same environment.
   - Competition for colonization sites and nutrients
   - Inhibition of pathogenic growth via the production of short-chain fatty acids (SCFA), antimicrobial substances, and bacteriocins
2. Indirect mode that improves the immunomodulatory actions of the body.
   - Modulation of immune responses by stimulating immune cells and activating cytokine and immunoglobulin production
   - Improved gut barrier integrity via mucin glycoprotein production
   - Providing energy to epithelial cells, improving villi growth, crypt development, and tight junctions
   - Interaction with the brain-gut axis.

Probiotics also regulate and restore the balance of the gut microbiota by stimulating the growth of beneficial bacteria and inhibiting the

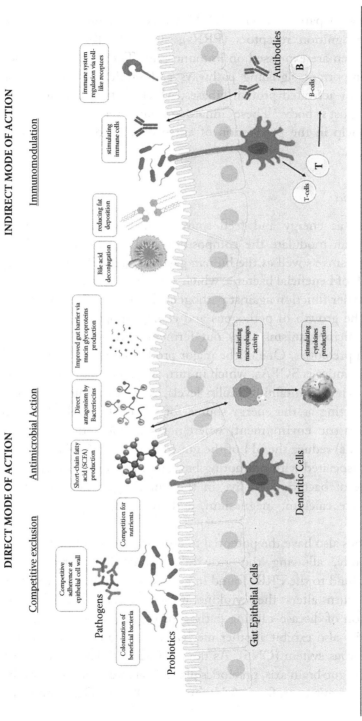

Figure 1.1 Mechanism of action of probiotics in humans in direct and indirect modes.

survivability of pathogens via SCFA. Moreover, probiotics identify pattern recognition receptors (PRRs), such as toll-like receptors (TLR), which are expressed on immune cells. Thereafter, probiotics regulate important signaling pathways and produce nuclear factor and mitogen-activated protein kinases, which in turn communicate with the host's body. These innate immune responses that get activated help in the production of anti-inflammatory cytokines or chemokines.

1.4.2 Prebiotics

By serving as energy and feed sources for the gut microbiome, prebiotics can modulate the composition and functioning of these microorganisms as well as the human gastric system. They stimulate the growth of beneficial bacteria, which are responsible for improving the gut barrier function against pathogens. Moreover, prebiotics also suppress the growth of pathogens and exhibit apoptosis simulation. The gut microorganisms carry out fermentation by acting on the starch and prebiotics. One of the major products from the fermentation of prebiotics is SCFAs, which in turn are associated with various functions, viz., maintaining inulin levels, regulating mucin expression, and acting as the energy source for the colon. Prebiotics also alter the gastric environment, wherein the fermentation products (mostly acids) reduce the pH of the gut from 6.5 to 5.5. This change in pH is associated with promoting butyrate formation by *Firmicutes* (a phylum of bacteria that live in the human gut), absorption of minerals like calcium, magnesium, and iron, and increasing bone density.

Prebiotics also have the potential to enhance the immune function of the body, allowing the growth of beneficial bacteria. The prebiotics bind to the PRRs found in gut-associated lymphoid tissue, which in turn alters the cytokine expression and decreases the proliferation of disease-causing pathogens (Figure 1.2).

Prebiotics also exhibit another mechanism to maintain the human central nervous system (CNS). As the GI tract is linked to the CNS through the gut-brain axis, prebiotics play a vital role in sustaining it by changing the gut microflora. Additionally, FOS and GOS modulate

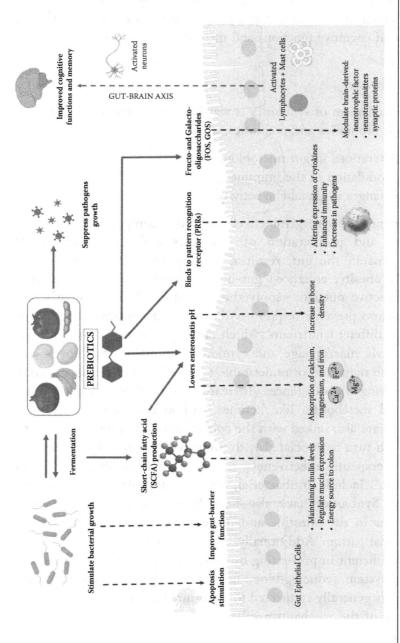

Figure 1.2 Mechanism of action of prebiotics in humans.

the brain-derived neurotrophic factor, neurotransmitters, and synaptic proteins. Prebiotics also allow the activation of lymphocytes and mast cells, which in turn are linked with the activation of neurons. Hence, improved cognitive functions and memory can be seen.

1.4.3 Synbiotics

The mechanism of synbiotics is via three major functions exhibited by both probiotics and prebiotics, namely:

1. Alterations in gut microbiome
2. Modulation of the immune system
3. Changes in metabolic activity

In general, the synergistic effect of the two components allows the growth and proliferation of probiotics and the inhibition of pathogens in the gut, resulting in improved digestion, a lower risk of obesity, enhanced gut-brain activity, reduced cholesterol, and effective nutrient supply throughout the body. Similarly, the dual action prevents respiratory diseases and improves the absorption of different nutrients, which altogether enhance the immune system via macrophage and cytokine stimulation. Synbiotics also allow the reduction of undesirable metabolites and the inactivation of carcinogenic substances. Increased levels of SCFAs and other required metabolites like ketones, carbon disulfides, and methyl acetates are also linked with the consumption of synbiotic products, which in turn is associated with improved host health. In terms of their therapeutic effectiveness, the most essential characteristics of synbiotics include antibacterial, anti-carcinogenic, and antiallergic effects. Synbiotics have also been known to counter the decay processes in the intestine and prevent the occurrence of diarrhea and constipation. Additionally, synbiotics have also proven to be highly efficient in preventing osteoporosis, regulating the immunological system, reducing blood fat and sugar levels, and curing brain disorders generally connected to abnormal hepatic functioning. The concept of the mechanism of action for synbiotics is presented in Figure 1.3.

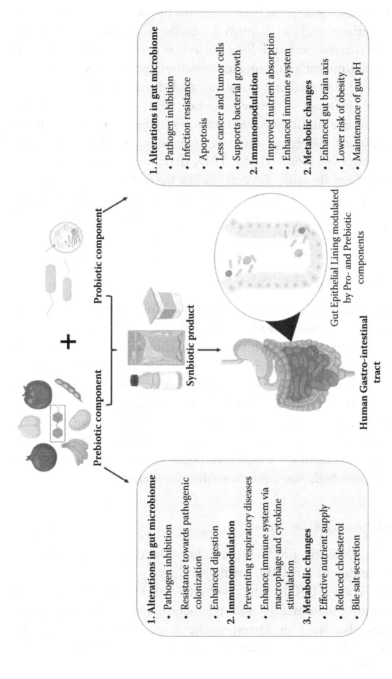

Figure 1.3 Mechanism of action of synbiotics in humans.

1.5 Health Benefits

Synbiotics have shown a vast role in humans with numerous beneficial effects such as improved hepatic functioning, enhanced growth of *Lactobacillus* and *Bifidobacterium* genera, preservation of the intestinal microbiota, enhanced immunomodulatory efficiency, prevention of bacterial translocation, and decreased occurrence of nosocomial infections. Recent studies and human trials have also shown the ability of synbiotics to either prevent or treat specific clinical disorders such as metabolic syndrome (MetS), irritable bowel syndrome (IBS), IBD, diarrhea, colon cancer, and kidney and liver diseases. Several studies have also investigated the potential of synbiotics in the reduction of mental-health-related conditions via the gut-brain axis (Marina González-Herrera et al., 2021).

1.5.1 Metabolic Syndrome

MetS, where the concerns are majorly related to obesity, is associated with the risk of coronary heart disease and type 2 diabetes. Numerous synbiotic formulations have been used in clinical trials to assess the improvement of MetS, wherein the synbiotic product has shown increased effectiveness of dietary treatment in the management of MetS and insulin resistance. MetS individuals have changes in their gut microbiome, including reduced diversity when compared to lean individuals. Thus, approaches employing the use of synbiotic (probiotic + prebiotic) supplementation allow gut microbiome modification, which has a positive effect on weight loss.

1.5.2 Inflammatory Bowel Disease

Numerous synbiotic products have been used in clinical studies for the treatment of IBD, namely, Crohn's disease (CD) and ulcerative colitis (UC). The synbiotic treatment allows for a reduction of inflammation and regeneration of epithelial tissue. Synbiotics also help in lowering the levels of TNF-α and its expression, which are associated with the pathogenesis of auto-immune diseases.

1.5.3 Irritable Bowel Syndrome

IBS is an intestinal disorder with major symptoms like intestinal pain, diarrhea, constipation, and bloating. The major reason for the occurrence of these symptoms is associated with an individual's genetics and gut microbiota, visceral hypersensitivity, changes in the brain-bowel axis, constant inflammation, and other environmental factors. Research has shown that synbiotic products containing inulin, *psyllium* fiber, *L. acidophilus*, *L. bulgaricus*, *L. plantarum*, *L. rhamnosus*, and *B. bifidum* significantly reduced the cases of chronic constipation in patients. The beneficial bacteria are essentially involved in preventing IBS by contributing to their proper growth in the presence of appropriate prebiotics. Their synergistic effects also help to protect the bowel epithelium by stimulating mucus production and enhancing intercellular tight junctions.

1.5.4 Diarrhea

Diarrheal diseases are associated with symptoms like loose or liquid bowel movements with increased frequency, water content, and volume, caused by infectious pathogens. There are various studies in the literature that describe the positive impact of synbiotic products on reduction of the duration of diarrhea along with intestinal mucosal healing. The use of synbiotic combinations containing *L. bulgaricus*, *B. animalis* subsp. *lactis*, and *S. thermophilus* as probiotics and inulin as prebiotics has been proven for their potential in preventing diarrhea, vomiting, and various infections in children.

1.5.5 Colon Cancer

Colorectal cancer is associated with the colon or the rectum of the human body and may be caused by diet, genes, or environmental attributes such as radiation and chemical carcinogens. Some studies using synbiotic combinations via *in vivo* models suggest that these treatments show therapeutic properties with beneficial effects such as reduced DNA damage in the colonic mucosa and reduced proliferation of pathogens. The introduction of synbiotic formulations in the diets has also shown significant changes in the gut microbiota when compared with the treatments with probiotics or prebiotics alone. Literature has shown certain clinical evidence depicting the potential of *L. rhamnosus, Bifidobacterium lactis,* and inulin in the reduction of the risk of colorectal cancer via cell proliferation inhibition, necrosis reduction, and maintenance of the integrity of the epithelial barrier. Similarly, prebiotics like FOS and GOS produce SCFA, which in turn stimulates a healthy bacterial population in the colon and provokes the protein and lipid metabolism to saccharolysis, thereby reducing carcinogenesis.

1.5.6 Kidney and Liver Diseases

Recent work on the treatment of chronic kidney and liver diseases has shown the potential of both probiotics and prebiotics, alone and in combination, as the potential solution for these conditions. A combination of FOS with *B. longum*, as a synbiotic formulation, has been used for treating liver disease. Similarly, prebiotics like lactitol and lactulose have also shown their potential to reduce the pH of the intestine by releasing H+ ions and converting ammonia into ammonium. This change leads to a concentration gradient, which in turn allows the reuptake of ammonia from the blood into the colon. Furthermore, the colon transit time also reduces with decreased ammonia content in the GI tract, which helps in treating hepatic encephalopathy (HE) in patients, which is characterized by liver cirrhosis and cognitive abnormalities.

1.5.7 Microbiome Gut–Brain Axis

In humans, mental health and immunity are inherently connected to one's gut health. Hence, a healthy gut can positively influence the function of other systems in the body. Researchers have investigated the ability of synbiotics to prevent mental health disorders such as Parkinson's disease or depression. The basic mechanism behind this effect suggests that intestinal microbes tend to release some chemical components that allow cell responses via the vagus nerve and send signals to the brain. This nerve connects the brain to the abdomen and acts as a modulator of the brain-gut axis in the human body. Prebiotics like GOS and FOS are reported to modulate brain-derived neurotrophic factors, neurotransmitters (d-serine), and synaptic proteins and regulate the neuroendocrine pathway. Nonetheless, prebiotics also control mood, learning, and memory, a few psychiatric disorders, by changing the composition of probiotics in the gut. Hence, the addition of both the probiotic and prebiotic components together as a synbiotic product on a regular basis can be employed for better physical, as well as mental health of the human body (Guiné et al., 2016).

1.6 Summary

- Probiotics, probiotics, and synbiotics exert positive effects on the host's gut health, metabolism, and immune system.
- The most common probiotics used at the commercial level include those from *Lactobacillus* and *Bifidobacterium* genera.
- The widely used prebiotics include inulin, FOS, GOS, and RFOs, which are obtained from major food groups like cereals, legumes, fruits, and vegetables.
- Synbiotic is a term used to describe a product that consists of a combination of probiotic(s) and prebiotic(s) that act synergistically and stimulate the growth and metabolism of intestinal microbes, thereby exerting beneficial effects on the host's health.
- Selection of appropriate prebiotics for a specific probiotic is a pre-requisite for formulating a synbiotic system in order to develop and maintain a good synergy between the two with maximum positive effects.
- Establishment of a basic mechanism of action for probiotics, prebiotics, and synbiotics is essential to design functional foods with enhanced quality to improve host health.
- The ability to control the composition of the probiotics and gut microbiota by prebiotic dietary substances is an interesting approach for the prevention and treatment of some major diseases in humans.

1.7 Multiple Choice Questions

1. Which one among the following is NOT a probiotic microorganism?
   a. *Lactobacillus rhamnosus*
   b. *Bacillus coagulans*

   c. *Saccharomyces boulardii*

   d. *Listeria monocytogenes*

2. Live beneficial microorganism enhancing gut microbiome is termed as _____.

   a. prebiotic

   b. synbiotic

   c. probiotic

   d. postbiotic

3. Prebiotics are the nutrients that are used by bacteria as a fuel source, and these include dietary fiber and _____.

   a. fat

   b. sodium

   c. iron

   d. carbohydrate

4. Which among the following is a common source of prebiotics?

   a. Chicory roots

   b. Strawberry

   c. Cabbage

   d. None of these

5. Prebiotics from plant sources are present in the form of _____.

   a. inulin

   b. FOS

   c. GOS

   d. All of the above

6. Prebiotics reduce the pH of the human gut from 6.5 to _____.

   a. 2.5

   b. 3.5

   c. 4.5

   d. 5.5

7. A prebiotic must be _____.

   a. resistant to stomach pH

   b. easily absorbed by gastric enzymes in the gastro-intestinal tract

   c. resist gut microbiota

   d. inactive at lower dosages

8. What is the minimum dose required for probiotic products when employed inside the food matrix?
   a. 10^5 CFU/mL
   b. 10^6 CFU/mL
   c. 10^7 CFU/mL
   d. 10^8 CFU/mL
9. Synbiotics are a combination of ____.
   a. probiotics and postbiotics
   b. prebiotics and probiotics
   c. prebiotics and psychobiotics
   d. probiotics and psychobiotics
10. Which among the following is true for the mechanism of action of probiotics?
   a. Provide colonization sites and nutrients
   b. Inhibit pathogen's growth via SCFA production
   c. Modulate immune response by killing immune cells
   d. Reduce gut barrier integrity

1.8 Short Answer Type Questions

Q1. Enlist the three major functions exhibited by probiotic and prebiotic components in the mechanism of synbiotics.
Q2. What are the beneficial effects exerted by synbiotics in preventing carcinogenesis?
Q3. What are the characteristics and properties of prebiotics?
Q4. Why the selection of an appropriate matrix is essential for the delivery of probiotics?
Q5. Enlist the important characteristics of a true 'probiotic.'

1.9 Descriptive Questions

Q1. Define probiotics, their mechanism of action, and the health benefits associated with them.
Q2. Enlist the difference between probiotics and prebiotics.
Q3. Write a short note on the following:
   a. Mechanism of action of prebiotics in the human gut
   b. Health benefits associated with synbiotics
   c. Sources of prebiotics

Q4. Describe the role of probiotics in the functioning of the gut-brain axis.

Q5. Describe in brief the major functions shown by probiotics and prebiotics in the mechanism of synbiotics.

1.10 Answers for MCQs

Q1	Q2	Q3	Q4	Q5	Q6	Q7	Q8	Q9	Q10
d	c	d	a	d	d	a	b	b	b

References

Adebola, O. O., Corcoran, O., & Morgan, W. A. (2014). Synbiotics: The impact of potential prebiotics inulin, lactulose, and lactobionic acid on the survival and growth of lactobacilli probiotics. *Journal of Functional Foods, 10*, 75–84. 10.1016/J.JFF.2014.05.010

Guiné, R., de, P. F., & Silva, A. C. F. (2016). Probiotics, prebiotics and synbiotics. In *Functional Foods: Sources, Health Effects, and Future Perspectives*. 10.1201/b15561-2

Marina González-Herrera, S., Bermúdez-Quiñones, G., Luz Ochoa-Martínez, A., Olga Rutiaga-Quiñones, M., & Gallegos-Infante, A. J. (2021). Synbiotics: A technological approach in food applications. *Journal of Food Science and Technology, 58*(3), 811–824. 10.1007/s13197-020-04532-0

Suggested Readings

Gibson, G. R., Hutkins, R., Sanders, M. E., Prescott, S. L., Reimer, R. A., Salminen, S. J., Scott, K., Stanton, C., Swanson, K. S., Cani, P. D., Verbeke, K., & Reid, G. (2017). Expert consensus document: The International Scientific Association for Probiotics and Prebiotics (ISAPP) consensus statement on the definition and scope of prebiotics. *Nature Reviews Gastroenterology and Hepatology, 14*(8), 491–502. 10.1038/nrgastro.2017.75

Hill, C., Guarner, F., Reid, G., Gibson, G. R., Merenstein, D. J., Pot, B., Morelli, L., Canani, R. B., Flint, H. J., Salminen, S., Calder, P. C., & Sanders, M. E. (2014). Expert consensus document: The international scientific association for probiotics and prebiotics consensus statement on the scope and appropriate use of the term probiotic. *Nature Reviews Gastroenterology and Hepatology, 11*(8), 506–514. 10.1038/nrgastro.2014.66

2

SYNBIOTICS IN DAIRY INDUSTRY

2.1 Dairy Products as Carriers of Probiotics and Prebiotics

For a decade, probiotic-based functional foods and supplements have shown increasing acceptability because of the numerous health benefits and nutritional well-being associated with their consumption. In the food industry, the dairy sector is the largest and fastest-growing sector for the utilization of probiotic and synbiotic products. The dairy industry allows the consumption of probiotics such as *Lactobacillus acidophilus*, *L. casei*, *L. rhamnosus*, *L. bulgaricus*, *Streptococcus thermophilus*, *Lactococcus lactis*, *Bifidobacterium bifidum*, and *B. longum* predominantly via products like fermented milk and yogurt, which are consumed regularly and are the most used probiotic carriers. Historically, fermented milk was the first probiotic food product, though, in recent times, different food matrices have been explored for delivering probiotics. Dairy-based food products are, in general, considered an important carrier for probiotics like *Lactobacillus* and *Bifidobacterium* species. This is because of the nutrition profile of dairy products, especially the presence of lactose, which is an essential requirement for the growth of these beneficial microbes in the human gut. Alongside, the addition of prebiotics such as inulin and fructooligosaccharides (FOS) improves the survivability of probiotics in dairy products and enhances their overall shelf life, which is again one of the main challenges encountered by dairy foods. In addition to this, the improved flavor and nutritional profile are the added benefits that allow better consumer acceptance.

Owing to the aforementioned advantages, the development of synbiotic dairy products is also growing in the food industry, wherein two major approaches have been identified for the efficient incorporation of probiotics into dairy products: (1) fermentation of the food item using selected probiotics, and (2) inoculation of probiotic strains into food matrices followed by digestion in the gastrointestinal (GI) tract.

DOI: 10.1201/9781003304104-2

Although probiotics are natural constituents of many fermented dairy products, for instance, yogurt, fermented dairy beverages, cheese, kefir, koumiss, sour cream, and cultured buttermilk, most of the fermented products do not contain well-defined probiotic strains. Hence, the recent trend in the production of pro/synbiotic dairy products or functional foods is either to carry out probiotic-based fermentation or to incorporate probiotic strains in nonfermented or fermented food products. Probiotics are either used alone or in combination with traditional starter cultures along with added prebiotics to create a synbiotic system with added health benefits. In the following sections, we will discuss various dairy products as potential probiotic delivery vehicles.

2.1.1 Milk

Milk is the most widely used carrier matrix for probiotics in the dairy industry owing to its high nutritional value, suitability for probiotic growth, consumer acceptability, and easy availability. Several probiotic milk products are widely used across the world. Even traditional delicacies like Kefir and Koumiss are probiotic-enriched fermented milk. Recent studies have attempted to improve human gut health and immunity by adding probiotic strains to traditional fermented milk-based products. Probiotics are also used in milk fermentation along with the traditional starter inoculum to improve the flavor profile as well as the physical characteristics of the product. For example, fermentation of milk by *L. plantarum* P-8 and/or *Lacticaseibacillus casei* DN-114001, along with yogurt starter cultures, *S. thermophilus*, and *L. delbrueckii* subsp. *bulgaricus* enhances the flavor of yogurt by producing flavor compounds like acetic acid, 3-methyl butanal, acetoin, caproic acid, butyric acid, and hexanal.

However, milk is not an ideal choice for many probiotic bacteria, such as lactic acid bacteria (LAB), for their growth and proliferation, thereby creating challenges for the production of probiotic dairy-based food products. Moreover, different probiotics exhibit different growth trends during the fermentation process. Thus, the selection of appropriate strains, along with added metabolites such as prebiotics, becomes essential for probiotic milk fermentation. Different techniques are followed in the dairy food industry to promote probiotic growth and get the desirable attributes required during fermentation. These are as follows:

- *Combining two or more microbial cultures*: Probiotics like *Bifidobacterium* species struggle to grow and survive in milk during the fermentation process, being obligate anaerobes and less proteolytic. Moreover, the optimum pH for the growth of Bifidobacteria is 6.5–7.0, and beyond this, the growth is retarded or inhibited. *Lactobacillus* can grow easily at a pH of 5.5–6.0. Thus, combining Bifidobacteria with *Lactobacilli* species allows their adequate growth. For example, *Lactococcus lactis* ssp. *lactis* amplifies the viability and subsequent growth of *B. longum* BB536 in fermented milk by protecting the latter against any active oxygen species and reducing the cases of oxygen damage.
- *Use of essential salts*: Salts like magnesium and calcium have shown their positive effects in encouraging the growth of *B. animalis* subsp. *lactis* Bb-12 in milk thereby enhancing the textural, sensorial, and physicochemical properties of the final product.
- *Addition of prebiotics*: Prebiotics are often added to milk and milk-based products to develop a synbiotic system and allow the growth of probiotics that cannot otherwise grow in a specific medium. In general, prebiotics like inulin and short-chain fructooligosaccharides (sc-FOS) are added in the form of their natural sources, such as fruit pulp (e.g., cornelian cherry, red grape, and black mulberry), to enhance the survival of probiotics in the product by serving as food to the probiotics. Moreover, the prebiotics also enhance the sensory acceptability of the product and allow scope for more options in fermented dairy items. Table 2.1 provides a summary of recent studies done on the formulation of synbiotic dairy-based food products.
- *Modification of redox potential*: The redox potential of the developed fermented milk or milk product can be modified as per the survivability of desirable probiotics by adjusting oxidation reduction using N_2 gas, which in turn results in significant growth of probiotics like *B. bifidum* without altering the fermentation kinetics of other strains.

Table 2.1 Summary of Recent Studies Done on the Formulation of Synbiotic Dairy-Based Food Products

SYNBIOTIC PRODUCT	PROBIOTIC	PREBIOTIC	FINDINGS	REFERENCES
Synbiotic fermented milk	*Lactobacillus rhamnosus* 4B15	Galactooligosaccharides (GOS)	• Antioxidant activity significantly increased by fermentation with the probiotic. • 39 peptides were detected, indicating the potential of GOS-enriched skim milk as a prebiotic substrate with the ability to prevent oxidative stress during production.	Oh et al. (2019)
Synbiotic fermented milk	*L. acidophilus* ATCC®4357™	Fructooligosaccharide (FOS) and isomaltooligosaccharide (IMO)	• Maximum growth of the probiotic (10^2 CFU/mL) was observed with 2.345–2.445% of FOS and 2.53–2.62% of IMO. • Significant antimicrobial activity was observed against *Escherichia coli* and *Staphylococcus aureus*, indicating significant therapeutic potential of the developed product.	Shafi et al. (2019)
Synbiotic yogurt	*Streptococcus thermophilus* and *L. bulgaricus*	Maltodextrin and gelatin matrix	• Fortification with encapsulated vitamin D_3 led to an increase in textural parameters where syneresis value decreased and viscosity increased. • Probiotic count was also in the desired range with sensory acceptance like control sample.	Nami et al. (2023)

(Continued)

Table 2.1 (Continued) Summary of Recent Studies Done on the Formulation of Synbiotic Dairy-Based Food Products

SYNBIOTIC PRODUCT	PROBIOTIC	PREBIOTIC	FINDINGS	REFERENCES
Synbiotic yogurt	L. acidophilus LA-5	Honey and cinnamon extract	• Rheological evaluation indicated that increasing additive's level made yogurt form a weak complex viscosity, in which cinnamon had superior influence compared to honey during the fermentation process and storage. • Microbial analysis showed that honey and aqueous cinnamon extract did not reduce the survival of probiotic during storage time.	Sohrabpour et al. (2021)
Synbiotic yogurt	Probiotic yeast Saccharomyces boulardii	Inulin	• Micro-rheological analysis showed that addition of inulin increased the solid properties of the synbiotic yogurt (0.582–0.595) compared with the plain yogurt (0.503–0.518) and reduced syneresis. • Addition of inulin improved the textural and sensory properties of the synbiotic yogurt, as well as survival of S. boulardii with viable count above 10^6 log CFU/g in yogurt.	Sarwar et al. (2019)

(Continued)

Table 2.1 (Continued) Summary of Recent Studies Done on the Formulation of Synbiotic Dairy-Based Food Products

SYNBIOTIC PRODUCT	PROBIOTIC	PREBIOTIC	FINDINGS	REFERENCES
Synbiotic milk powder (SMP)	L. plantarum JIBYG12	Inulin (INU), xylo-oligosaccharide (XOS), fructooligosaccharide (FOS), and iso-malto-oligosaccharide (IMO)	• Adding 1.2% IMO, the probiotic count and the calcium enrichment increased by 8.1% and 94.5% compared with those of the control group. • In SMP calcium absorption, serum calcium and phosphorus levels increased, indicating positive effect of calcium supplementation capacity to improve bone health in SMP.	Jia et al. (2023)
Synbiotic freeze-dried yogurt powder	L. plantarum	Sorbitol	• Probiotics in yogurt powders containing microcapsules enriched with the sorbitol were higher. • Synbiotic powders also had high solubility and the ability to produce reconstituted yogurt with the same viscosity and color as fresh yogurt.	(Jouki et al., 2021)
Synbiotic ice cream	L. acidophilus LA-5	Camel milk and black rice powder (BRP)	• Adding BRP to ice cream blends resulted in significant increases in the overrun, viscosity, melting resistance, and probiotic viability in ice cream samples. • Formulation with 25%–50% of BRP had the most accepted sensory scores.	Elkot et al. (2022)

Moreover, along with the aforementioned methods, control over the processing and storage temperatures is also crucial in determining the viability of probiotics in fermented dairy products as per the recommended standards of 10^6 to 10^7 CFU/g or CFU/mL.

2.1.2 Yogurt

After fermented milk, yogurt is the next most used functional food rich in probiotics. A typical yogurt is made by the fermentation of milk using probiotic strains *L. delbrueckii* subsp. *bulgaricus* and *S. thermophilus*. However, recent studies have also explored the potential of other probiotic strains in producing probiotic yogurts with various probiotic strains that are either used for fermentation or incorporated into yogurts. Additionally, an increased need for a new range of dairy products, like those containing a blend of both probiotics and prebiotics, has led to the development of functional foods like synbiotic yogurts at the commercial level.

Synbiotics are known to influence lipid profiles and protect the human body against colorectal cancer than probiotics or prebiotics alone. Hence, such a synergism is essential for the associated health benefits to the host. Moreover, the prebiotics, if provided from natural sources (like fruit or vegetable-based yogurt), not just enhance the overall probiotic colonization by serving as a delivery matrix but also provide exceptional flavor to the yogurt. Other prebiotics, such as inulin, have also shown a positive effect on the bacterial community in fermented milk (Table 2.1). Inulin is also used to make better-textured low-fat synbiotic yogurt, along with the health-promoting strains of *Lactobacilli* and *Bifidobacteria*.

2.1.3 Cheese

Among the different dairy matrices for the delivery of probiotics to the GI, cheese has also been explored in recent studies to prove its efficiency. Cheese can be considered a good carrier of probiotics because of its (1) low water activity ($a_w > 0.90$); (2) low storage temperature (4–8°C); (3) high pH (4.8–5.6) (in comparison to synbiotic fermented milk and yogurt (3.7–4.5)); and (4) suitable texture properties (such as high-fat percentage, which tends to protect probiotic microbes during the gastric transit). Nonetheless, two major criteria need to be taken care of while adding probiotics to cheese. Firstly, the probiotic strain must be able to survive and flourish during the steps of cheese making, namely, salting, ripening, and storage. Secondly, the probiotics and externally added prebiotics must not alter the sensory profile of cheese in a negative way. Probiotics (*Bifidobacterium* and *Lactobacillus* genera) and prebiotics (inulin, FOS, and polydextrose) are added to cheese to enhance the nutritional composition and increase consumer acceptability toward healthier dietary options. In the cheese manufacturing process, during the steps of cooking and cheddaring, the presence of metabolically active LAB can compete with added probiotics for the prebiotics in the cheese, affecting their overall growth and compromising the desirable viable count. Thus, the subsequent stages, like salting and milling, are considered for probiotic inoculation in cheese in order to maintain the recommended probiotic count of $>10^7$ CFU/g.

However, several researchers have also stated the negative influence of probiotics on the sensory profile of cheese. For that purpose, encapsulated probiotics are being added to cheese to preserve the flavoring compounds and metabolites to exhibit desirable sensory probiotics in cheese. The synbiotic cheese has gained more scores in the overall acceptability of consumers in comparison to the basic cheese owing to the better sensory profile, creaminess, more free fatty acid content, and added health benefits (Table 2.1).

2.1.4 Ice Cream

Ice creams are also one of the dairy-based options for carrying probiotics, and it is highly recommended to make ice creams a functional healthy product. Moreover, the viability of probiotics during the processing and storage is higher in comparison to many other dairy items, making ice creams a potential probiotic delivery matrix. Nonetheless, there still exist a few technological factors that tend to alter probiotic growth and viability. The commercial industry also demands optimum methods to maintain the stability between probiotic survivability and the physicochemical attributes of ice creams.

The general steps of unit operations in the manufacturing process of the synbiotic ice cream include raw material procurement, base mixing, homogenization and pasteurization, aging, fermentation, incubation, cooling, flavor addition and continuous freezing, packaging, and finally hardening. The process parameters that affect the probiotic growth in ice creams include the fermentation process (time, temperature, and probiotic inoculation), churning and freezing, overrun process, storage time and conditions, and thawing of ice cream. For the fermentation process, the literature suggests that fermenting ice cream with probiotics is much better than adding the freeze-dried probiotic culture to the ice cream mix without any subsequent fermentation. Several studies support the fact that when an ice cream mix is fermented with probiotics, the required count of 10^6 CFU/g for the probiotic strain is maintained for a longer period. On the other hand, ice creams with probiotics added directly to the mix, without any fermentation process, have viable counts for a comparatively shorter duration. Additionally, ice cream mixes should be fermented and frozen under regulated conditions with a pH of around 5.6 for better probiotic viability because a lower pH can also decrease the growth of probiotics. Apart from this, the two most important steps in ice cream making, freezing and churning, have the greatest impact on the viability of probiotics, wherein the count significantly decreases after these processes. Similarly, overrun levels, or the step of adding air to ice creams for desirable texture, also affect the probiotic count in the final developed ice creams. In general, lower overrun levels are associated with a desirable probiotic count in the ice cream supplemented with any prebiotic source.

Thus, the addition of probiotics has a positive impact on making ice cream with better quality characteristics. Nonetheless, the addition of prebiotic components like inulin and oligofructose can also improve the overall profile of such products with added health benefits, different flavors, and desirable probiotic counts to make synbiotic ice cream.

2.1.5 Butter and Cream

Other dairy products that can be used as potential probiotic carriers include butter and cream. The incorporation of probiotics in such products is useful as they have widespread recognition worldwide. Butter is mainly composed of fats and though it has many health benefits, consumers nowadays prefer healthier versions or substitutes of butter and similar products. Hence, the application of selective probiotics and prebiotics into butter can be used to make novel synbiotic butter that can exhibit itself as a better and healthier functional food candidate. Several studies have shown that probiotic bacteria like *Lacticaseibacillus casei* subsp. *casei* and *L. maltaramicus* reduce the cholesterol content of butter and cream. Moreover, fermentation of cream with probiotics like *B. bifidum, S. thermophilus, L. bulgaricus,* and *L. acidophilus* generates short-chain fatty acids (SCFAs) like capric, butyric, and caproic acids when supplemented with plant oils (hazelnut oil, sunflower oil, and soybean oil). Thus, the incorporation of prebiotics like inulin together with probiotics can be used as an additional benefit that can upsurge the demand for synbiotic-based dairy products in the commercial market. Prebiotics can be added in dry form either before the churning or ripening step or at the end, along with salt, before packaging and storage.

2.1.6 Powdered Milk and Infant Formulas

Though the addition of anything extra to infant feed remains a topic of concern and goes through numerous scrutiny factors, the European Society for Pediatric Gastroenterology, Hepatology, and Nutrition has stated that probiotic/prebiotic-supplemented infant formulae do not raise any health concerns about the growth of an infant. Additionally, the motive for adding probiotics to infant formula is to alter the equilibrium of intestinal microbiome in infants in such a way that it helps in the development of an efficient immune system. In recent times, infant food mixes and medicines are supplemented with probiotics (*lactobacilli* and *bifidobacteria* strains), and prebiotics, and their use has become more prevalent in neonatology. Several studies have reported a positive impact of synbiotic infant powders with desired and improved results in comparison to breast milk and sterile water.

Nonetheless, other powder-based synbiotic dairy food products include milk powder, ice cream, and yogurt mixes. Such products that have gone through the drying steps via methods like spray drying and freeze drying are more effective in preserving probiotics and increasing the overall shelf life of the product. Spray drying is often accompanied by encapsulation material, such as starches, gums, maltodextrin, inulin, and whey proteins, which also acts as an additional prebiotic to create a synbiotic system in powder dairy products. The encapsulation technique also plays a crucial role in maintaining the stability of probiotics during the spray-drying process (Table 2.1).

2.2 Health Benefits

Probiotic strains in dairy items, along with prebiotics, exhibit a plethora of health benefits and physiological effects in the form of a synbiotic functional food product that is supported by various *in vivo* and clinical-level research.

2.2.1 Milk

There are various health-promoting advantages of milk fermented by probiotic strains in the presence of prebiotics in a synbiotic system.

- Synbiotic-fermented milk contains various bioactive peptides associated with different health benefits. For instance, milk fermented by *Lacticaseibacillus casei* is known to enhance the regulation of immunoglobulin A (IgA), interleukin-6 (IL-6), and IL-10 levels and maintain intestinal health and mucosal cellular immunity.
- Synbiotic milk fermented by *B. animalis* subsp. *Lactis, L. delbrueckii* subsp. *bulgaricus, Lactococcus lactis* subsp. *cremoris,* and *S. thermophilus* induces a decrease in intestinal pH, modifies the SCFAs profile, and increases the abundance of beneficial gut microbes, thereby inhibiting the growth and survival of *Enterobacteriaceae* strains that are responsible for causing ulcerative colitis.
- Fermented synbiotic milk containing *L. acidophilus* La-5 and *B. bifidum* BB-12 alleviates the occurrence of gastric and lower abdominal pain and cases of irritable bowel syndrome (IBS) in adults.

- Probiotic-fermented milk supplemented with prebiotics also helps in minimizing blood serum cholesterol and low-density lipoprotein cholesterol levels and normalizing cases of hypertension or mild blood pressure. Such products also allow the regulation of lipid concentrations.

2.2.2 Yogurt

- Synbiotic yogurts have shown numerous health-promoting properties as a result of their high antioxidant activity, angiotensin-converting enzyme (ACE) inhibitory activity, and reactive oxygen species (ROS) scavenging activity. These benefits include antidiabetic and antihypertensive activities along with immunomodulatory properties.
- Synbiotic yogurts with *L. acidophilus* and FOS exhibit hypolipidemic potential with the ability to improve hepatic steatosis and liver enzyme concentrations. Moreover, such a product is also effective in improving cardiovascular and metabolic risk factors in humans.
- Synbiotic yogurt with probiotics *L. acidophilus* La-5 and *B. bifidum* BB-12 also suppress *Helicobacter pylori* infection by reducing the *Streptococcus* strains present in saliva and gut and increasing the output of IgA, thereby preventing GI and lower respiratory tract infections.
- Yogurt made using the strain of *Bifidobacterium* also decreases the chances of gut mutagenicity by enhancing the intestinal microbial diversity, which in turn is associated with the activation of T helper cells and cytokines. Additionally, inflammation in the gut also decreases, which is associated with the inhibition of anti-inflammatory metabolite production in the gut.

2.2.3 Other Products

Besides the synbiotic milk and yogurts, there are only a few health claims related to the use of products like synbiotic cheese, infant formulas, and ice cream. For these products, the addition of probiotics and prebiotics is linked more to the physicochemical and sensory profiles of the product. However, equal importance should be given to the compatibility between prebiotics and probiotics. Nonetheless, few studies have shown that supplementation with probiotics and pre-biotics tends to modulate the colonic microbiome in a positive way. The combination also increases the metabolites such as propionate, lactic acid, acetate, and butyrate required for the growth and proliferation of beneficial gut microbes. Moreover, competitive exclusion by probiotics toward pathogens like *Salmonella* in the intestinal gut lining has also been observed, indicating the associated health benefits like maintenance of the immune response system, modulation of the gut microbiome, and reduction of cholesterol levels (Figure 2.1).

2.3 Challenges Faced by Synbiotic Dairy Products

The maintenance of desirable and recommended viable count of probiotics is an essential criterion to claim a product as probiotic and synbiotic, failing which the efficacy of probiotic products loses their value. In this regard, the synbiotic dairy-based food industry faces several challenges that need to be dealt with efficiently. In the manufacturing process, probiotic microorganisms come across various stressful and harsh conditions during food processing, storage, and while passage through the GI of the human body after consumption. As per different international standards, the minimum viable count for probiotics at the time of consumption should be 10^6 CFU/g (for *B. bifidum*) and 10^7 CFU/g (for *L. acidophilus*) in probiotic-fermented milk.

As discussed previously, some of the probiotic strains like *Bifidobacteria* are oxygen-sensitive, and their viability gets significantly altered by the oxygen present in the synbiotic dairy products. This eventually leads to toxicity in such products which subjects itself as a significant technological hurdle in the formulation of synbiotic-fermented milk, yogurt, milk powder, ice cream, etc. The species of *Bifidobacterium* has an intestinal source with anaerobic metabolism and their exclusive dependence on metabolism via fermentation. The growth of such strains in the presence of oxygen causes oxygen toxicity, which is further associated with a significant decay in the viable count of *Bifidobacteria* in synbiotic dairy-based food products. The interaction of these probiotic strains to dissolved oxygen in gut

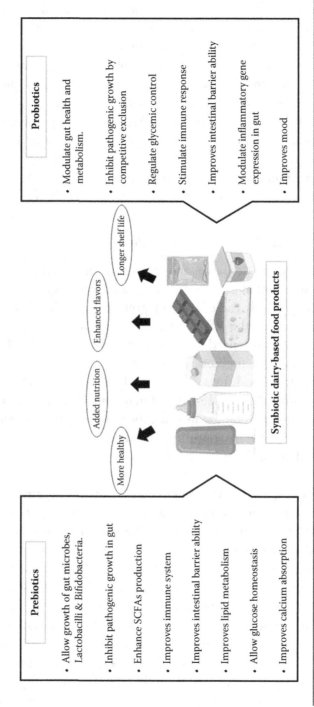

Figure 2.1   Health benefits of synbiotic dairy-based food products in humans.

causes the adherence of toxic oxygenic metabolites at the intracellular level which eventually causes their death. Apart from this, at the various stages of processing dairy products such as during the mixing steps, the inevitable addition of oxygen occurs in the product. Additionally, during the storage of such products, certain packaging materials allow the movement of atmospheric oxygen into the product. In this way, an eventual decrease in the probiotics occurs, reducing its overall efficiency.

Apart from oxygen, the low pH of different dairy items also negatively influences the survivability and growth of the probiotic component of the synbiotic product.

2.3.1 Possible Solutions for These Challenges

Irrespective of the challenges faced by the synbiotic dairy industry, there exist possible solutions and techniques to overcome the insufficient probiotic viability and their subsequent death. Firstly, the selection of appropriate probiotic strains with oxygen-resistant properties should be carried out at the preliminary level. In addition, a selected probiotic strain must also tolerate the harsh acid and bile stresses of gastric environment as survival under such conditions tends to improve stress tolerance as well as the biological properties of *Bifidobacterium* species.

Along with this, oxygen, which can cause severe oxygen toxicity, can be removed from the product at the molecular level by using products such as the following:

1. oxygen-scavenging substances such as natural antioxidants (tocopherols) and synthetic antioxidants (butylated hydroxytoluene and butylated hydroxyanisole),
2. containers with passive barrier properties (multilayered packages with aluminum foils),
3. packaging materials with oxygen impermeability.

In addition to these, encapsulation of probiotics and addition of nutrients like prebiotics to dairy products can improve and maintain probiotic viability throughout the product storage life.

Microencapsulation is a widely used technique because of its cost-effective and sustainable technology that can be used easily to protect probiotics from various processing steps as well as internal gut conditions and allow adequate viable probiotic count. Numerous studies have concluded positive effects of microencapsulated probiotics to achieve higher survival rates in comparison to the non-encapsulated ones. The various encapsulating materials used for dairy-based probiotics include those comprising chitosan like alginate–chitosan and chitosan/dextran sulfate; prebiotics like oligosaccharides and inulin; polyols such as sugar and alcohol fibers; gelatin; gums such as xanthan gum, gum Arabic, and maltodextrin; carrageenan; alginate-based materials like sodium alginate, alginate–$CaCO_3$, calcium alginate, alginate, and human-like collagen; cellulose derivatives; vegetable proteins; pectin hydrogel beads; milk protein-based substances such as whey protein isolates, and many more. The different encapsulation methods include extrusion, emulsification, freeze drying, spray drying, spray chilling, electro-hydrodynamic atomization, coacervation, and gelation. Apart from these, the use of recent technologies like embedding techniques (probiotic microencapsulation via microgels or microcapsules) and electrospinning (adding probiotics in nanofibers) is also employed as a part of microencapsulation to ensure effective long-term stability and viability of probiotics in synbiotic dairy products. While the use of these new innovative techniques has shown efficient results, conventional microencapsulation methods like spray drying are still the most widely used technique at the commercial level used by the synbiotic dairy industry.

As discussed in the earlier sections, the addition of several salts and prebiotic ingredients along with cryoprotectants also helps in protecting and improving the growth and survivability of probiotics in dairy products during their storage life. Just like most of the encapsulation techniques, cryoprotectants also safeguard probiotic microorganisms from harsh freeze-drying conditions. The commonly used cryoprotectants and prebiotics for maintaining the probiotic count in various synbiotic dairy products like yogurt include poly-γ-glutamic acid, soluble corn fiber, chicory-inulin, polydextrose oligofructose, and inulin. Such agents not just help to maintain the count of probiotics but also sustain the textural and rheological properties of the dairy product. In addition to these methods, immobilization of probiotic strains by natural fruits and berries as the immobilization carrier in synbiotic dairy products has also been reported in recent studies as a suitable technique to enhance the probiotic growth during processing, storage, and the gastrointestinal transfer. Similarly, the use of natural antioxidant-rich fruit pulps such as soursop, custard apple, and sweetsop is also being used as a potential prebiotic agent to significantly improve the survival of specific probiotic strains in synbiotic dairy items.

Hence, in this way, smart application of novel techniques during the processing and storage of the developed product as well as after ingestion in the GI tract is important to ensure the sustenance of probiotic strains in a synbiotic dairy product along with the desired physicochemical and sensory acceptance of the product.

2.4 Summary

- Due to abundance of numerous health-promoting effects connected with the consumption of synbiotics in humans, such products are now widely accepted by consumers globally.
- Synbiotics in the dairy sector have been used to make probiotic-fermented prebiotic-enriched food products, dietary supplements, pharmaceuticals, and related organic therapeutic ingredients.
- Such products not just play a crucial role in attaining nutritional well-being and economic development but also provide dietary options to the health-conscious consumers, in the most simplified way.
- The dairy industry being the largest and fastest-growing sector in the food and beverage industry allows the scope for maximum utilization of synbiotic dairy-based products as potential probiotic carriers enriched with prebiotics.
- Several synbiotic products such as fermented milk, powdered instant milk, cheese, ice cream, yogurt, and butter have been clinically tested, and scientifically approved, and are now available in different parts of the world at the commercial level.
- The utilization of novel emerging techniques like encapsulation and the use of cryoprotectants and essential salts can help stabilize and formulate a synbiotic dairy-based product with desirable characteristics.

• Exploring other associated innovations like paraprobiotics and postbiotics, either alone or in combination with synbiotics, can be employed to develop novel functional food products with added health benefits.

2.5 Multiple Choice Questions

1. Which of the following traditional delicacies are probiotic-enriched fermented milk?
    a. Sauerkraut and Kimchi
    b. Kefir and Kimchi
    c. Kefir and Koumiss
    d. All of the above

2. Spray drying is often accompanied by ____ to create a synbiotic system in powder dairy products with enhanced shelf life.
    a. nanotechnology
    b. encapsulation
    c. freeze drying
    d. preservatives

3. Probiotics are added to cheese during ____. (Multiple answers)
    a. cheddaring
    b. cooking
    c. salting
    d. milling

4. Steps in ice cream making that significantly reduce the probiotic count is/are ___. (Multiple answers)
    a. freezing
    b. fermentation
    c. churning
    d. overun

5. Fermentation of cream with probiotics generates SCFAs like ____.
    a. capric acid
    b. butyric acid
    c. caproic acids
    d. All of these

6. Which one of the following probiotics can reduce the cases of irritable bowel syndrome (IBS) in adults?
   a. *L. acidophilus* La-5
   b. *B. bifidum* BB-12
   c. Both a and b
   d. None of the above

7. Ice cream mix should be fermented and frozen under regulated conditions with a pH of around ____ for better probiotic viability.
   a. 4.6
   b. 5.6
   c. 2.6
   d. 1.6

8. The major challenges faced by synbiotic dairy industry are related to _____.
   a. oxygen and pH
   b. water and oxygen
   c. Both a and b
   d. None of the above

9. Which one of the following operations is used for the incorporation of probiotics in nanofibers?
   a. Embedding techniques
   b. Electrospinning
   c. Cryoprotectants
   d. Extrusion

10. Which one of the following are the most commonly used cryoprotectants used in the synbiotic dairy industry?
    a. Poly-γ-glutamic acid, soluble corn fiber
    b. Chicoryinulin, polydextrose oligofructose
    c. Inulin, prebiotics
    d. All of the above

2.6 Short Answer Type Questions

Q1. How do dairy products exhibit themselves as carriers of probiotics and prebiotics?

Q2. Which are the most widely used synbiotic dairy food products?

Q3. Highlight the health benefits associated with synbiotic products other than milk and yogurt.

Q4. What are the challenges faced by the industry while manufacturing the synbiotic dairy products?

Q5. Enlist the possible solutions for the challenges faced by synbiotic industry and discuss any one in detail.

2.7 Descriptive Questions

Q1. Discuss in detail 'dairy products as carriers of probiotics and prebiotics.'

Q2. Enlist the different synbiotic dairy products used in the food industry. Elaborate about the synbiotic efficacy of two of those products.

Q3. What are the health benefits associated with different synbiotic dairy items?

Q4. Discuss the manufacturing steps of a synbiotic dairy food product while mentioning the critical control points.

Q5. What are the major challenges faced by synbiotic dairy industry? What are the possible solutions for these challenges?

2.8 Answers for MCQs

Q1	Q2	Q3	Q4	Q5	Q6	Q7	Q8	Q9	Q10
c	b	c, d	a, c	d	c	b	a	b	d

References

Elkot, W. F., Ateteallah, A. H., Al-Moalem, M. H., Shahein, M. R., Alblihed, M. A., Abdo, W., & Elmahallawy, E. K. (2022). Functional, physicochemical, rheological, microbiological, and organoleptic properties of synbiotic ice cream produced from camel milk using black rice powder and *Lactobacillus acidophilus* LA-5. *Fermentation, 8*(4), 187. 10.3390/FERMENTATION8040187

Jia, M., Luo, J., Gao, B., Huangfu, Y., Bao, Y., Li, D., & Jiang, S. (2023). Preparation of synbiotic milk powder and its effect on calcium absorption and the bone microstructure in calcium deficient mice. *Food and Function, 14*(7), 3092–3106. 10.1039/d2fo04092a

Jouki, M., Khazaei, N., Rezaei, F., & Taghavian-Saeid, R. (2021). Production of synbiotic freeze-dried yoghurt powder using microencapsulation and cryopreservation of *L. plantarum* in alginate-skim milk microcapsules. *International Dairy Journal*, *122*, 105133. 10.1016/ J.IDAIRYJ.2021.105133

Nami, B., Tofighi, M., Molaveisi, M., Mahmoodan, A., & Dehnad, D. (2023). Gelatin-maltodextrin microcapsules as carriers of vitamin D3 improve textural properties of synbiotic yogurt and extend its probiotics survival. *Food Bioscience*, *53*. 10.1016/j.fbio.2023.102524

Oh, N. S., Kim, K., Oh, S., & Kim, Y. (2019). Enhanced production of galactooligosaccharides enriched skim milk and applied to potentially synbiotic fermented milk with *Lactobacillus rhamnosus* 4B15. *Food Science of Animal Resources*, *39*(5). 10.5851/kosfa.2019.e55

Sarwar, A., Aziz, T., Al-Dalali, S., Zhao, X., Zhang, J., Ud Din, J., Chen, C., Cao, Y., & Yang, Z. (2019). Physicochemical and microbiological properties of synbiotic yogurt made with probiotic yeast *saccharomyces boulardii* in combination with inulin. *Foods*, *8*(10). 10.3390/foods81 00468

Shafi, A., Naeem Raja, H., Farooq, U., Akram, K., Hayat, Z., Naz, A., & Nadeem, H. R. (2019). Antimicrobial and antidiabetic potential of synbiotic fermented milk: A functional dairy product. *International Journal of Dairy Technology*, *72*(1). 10.1111/1471-0307.12555

Sohrabpour, S., Rezazadeh Bari, M., Alizadeh, M., & Amiri, S. (2021). Investigation of the rheological, microbial, and physicochemical properties of developed synbiotic yogurt containing *Lactobacillus acidophilus* LA-5, honey, and cinnamon extract. *Journal of Food Processing and Preservation*, *45*(4). 10.1111/jfpp.15323

3

SYNBIOTICS IN NON-DAIRY INDUSTRY

The dairy products are the most suitable and widely explored matrices for the delivery of probiotics and the development of synbiotic food products because of their highly nutritive composition. However, other food groups, such as fruits and vegetables, are also major food groups used for the same purpose. Even in the past few years, other food groups, such as cereals and pulses, have also shown their potential as a prebiotic matrix, which can be fermented using various probiotics to produce a non-dairy synbiotic food product. The consistent acceptance of non-dairy synbiotic products by consumers is primarily because of the shortcomings exerted by synbiotic dairy products, such as their limited shelf life, unacceptability by people with conditions like lactose intolerance, cholesterol-related issues, milk allergies, and even inclination toward veganism. These concerns have led to the growth of synbiotics and non-dairy-based food products in the functional food market. Synbiotic products developed using plants, fruits, vegetables, cereals, and pulses do not have cholesterol or lactose, preventing any chances of lactose intolerance or allergies to milk proteins (Chaturvedi, Chakraborty, 2021). Moreover, such non-dairy synbiotic foods are also rich in carbohydrates, protein, vitamins, minerals, fiber, antioxidants, and several other beneficial compounds, thus allowing the body to remain active, and nutritionally fulfilled, with the ability to repair and control cells.

Another important aspect associated with the development of such products is the added nutrition from different sources, which include:

1. the prebiotic matrix itself,
2. the numerous health benefits associated with probiotics,
3. the postbiotics produced after the fermentation process, and

DOI: 10.1201/9781003304104-3

4. the synergistic effect exhibited by the combination of the probiotic microbe and the prebiotic food product.

Over the past few years, plant-based and non-dairy synbiotic food products have found their way into the daily dietary patterns of health-conscious as well as ordinary people in the form of different ready-to-eat convenience food items like snacks and drinks. Several studies have shown that non-dairy synbiotic products are shelf-stable in comparison to their dairy counterparts when manufactured using food matrices like water, sugar, tea, and bakery ingredients as a source of prebiotics along with various recent technological advancements (Kumar et al., 2022). The commonly used probiotics in the functional food industry and even in the non-dairy industry belong to the family of *Lactobacillus*, followed by *Bacillus* for use in many commercial products. These microbes are now used in the form of their spores because of their thermophilic nature and thereby help tolerate the high processing temperatures. Hence, the selection of a suitable prebiotic matrix with respect to a specific probiotic microorganism became the priority while developing non-dairy synbiotic food products.

3.1 Need for Non-Dairy Synbiotics

Traditionally, dairy products were considered the best carriers for the delivery of probiotic microorganisms. However, several limitations associated with dairy-based probiotic and synbiotic food products induced the need for non-dairy synbiotic interventions. One of the major problems associated with dairy products is their unacceptability by lactose-intolerant individuals. Hypolactasia, lactose malabsorption, or lactose intolerance are caused by a deficiency of an enzyme present in the human body called 'lactase.' This enzyme is responsible for the breakdown and further digestion of the sugar 'lactose' in the small intestine. Hence, an individual suffering from lactose intolerance cannot digest lactose, which in turn causes symptoms like bloating, abdominal cramps, stomach pain, flatulence, and diarrhea. These symptoms occur as a result of colonic fermentation of unabsorbed lactose by the gut microflora, leading to the production of short-chain fatty acids (SCFA), methane, hydrogen, and carbon dioxide, thus

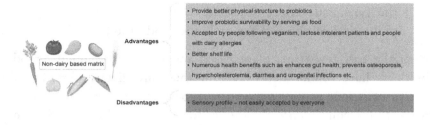

Figure 3.1   Advantages and disadvantages of different non-dairy food matrices for the development of synbiotic food products.

increasing time for gastric emptying and intracolonic pressure. Since lactose is a naturally occurring sugar in mammalian milk and dairy products, the consumption of such products becomes troublesome for individuals suffering from lactose intolerance. The severity of the symptoms varies from one person to another as the degree of intolerance remains individual-specific. In addition, this depends on the amount of lactose that has been taken by the individual. Experts estimate that about 68% of the world's population has lactose malabsorption, and the possible treatments for such conditions only include diet modifications. Moreover, a section of the world's population also belongs to individuals who follow concepts like veganism and are health-conscious. Such individuals, too, do not eat dairy products as per their own will and beliefs. However a diet without dairy can cause health issues like reduced calcium intake and weaker bones. Hence, synbiotic non-dairy food products can serve as a potential substitute and dietary option for such a population with required nutrition.

For that purpose, the unique physiology of plant-based non-dairy food products not just provides required nutrition as such, but the combination of probiotics also imparts added health benefits, as shown in Figure 3.1. These include reduced oxidative stresses and controlled glycemic index; anti-cancer and antimicrobial properties; and enhanced SCFAs and bioactive production, as well as improved gut health.

3.2 Non-Dairy Products as Potential Probiotic Carriers

In the non-dairy industry, fruit and vegetable juices are the most common and widely accepted probiotic products in the market, followed by jellies, jams, and pastes. Other than juices, non-dairy probiotic or synbiotic products are less explored at the commercial level. However, a lot of scientific studies identified non-dairy synbiotic ingredients, but not on a commercial scale (Mishra et al., 2021). These include probiotic cereal, legumes, and millet-based beverages, either individually or in combination. The recent application of cereals and legumes has proven that these food groups can also be explored as potential non-dairy matrices fulfilling the required criteria of probiotic growth as well as added health benefits and acceptance by consumers at the same time.

In general, for a product to be used as a potential probiotic carrier, it must have the following characteristics:

1. It should allow the growth of probiotics during normal as well as processing and storage conditions.
2. It must provide a matrix to tolerate the harsh unfavorable conditions of the gut.
3. It should be able to sustain and thrive for a longer duration to provide a better shelf life to the food product.

The potential of different non-dairy food groups as probiotic vehicles has been discussed separately below:

3.2.1 Fruits and Vegetables

In the fruit sector, a lot of fruit juices that can be used as a medium for the delivery of probiotic microorganisms have been explored and are even readily available in the market. These include apples, mangoes, oranges, grapes, cranberries, sweet lime, grapes, pineapple, strawberry, carrots, pomegranate, prickly pear, and beetroot. Fruits are a rich source of nutrients such as vitamins, minerals, dietary fibers, antioxidants, and prebiotics (fructooligosaccharides [FOS] and raffinose-family oligosaccharides [RFOs]). Similarly, a lot of vegetables like bamboo shoots, potato, broccoli, eggplant, mustard leaves, tomato, onion, garlic, peanuts, ginger, cabbage, and cauliflower have been researched and proven as a suitable non-dairy matrix with the ability to allow the growth of various probiotic microorganisms. Vegetables are also rich in vitamins, minerals, phytonutrients, phytochemicals, and carbohydrates. They help maintain the body's structure and facilitate maintaining alkaline homeostasis within the body.

Several vegetables like onions, garlic, and mushrooms are considered the best sources of prebiotics like inulin and FOS. Moreover, vegetables like cabbage and beetroot have been used to make traditional fermented probiotic dishes like *kombucha, kimchi*, and *soido*.

The main probiotic microorganisms used for the development of fermented foods using fruits and vegetables belong to the species of *Lactobacillus* and *Bifidobacteria*, such as *L. plantarum, L. casei, L. paracasei, L. acidophilus, L. rhamnosus, L. fermentum,* and *Bifidobacterium bifidum*. Additionally, some factors also impact the stability of probiotics in fruit and vegetable matrices. These include (a) the type of probiotic strain and its inoculum size; (b) the nutritional composition of the fruit or vegetable, i.e., water activity, pH, oxygen concentration, and acidity; and (c) the production, packaging, and consequent handling of the final product.

Thus, fruits and vegetables can be considered suitable substrates for developing synbiotic microorganisms, primarily because of their rich nutritional composition. They serve as a source of prebiotics and can be fermented easily with a flavor profile that appeals to people of all ages and are regarded as a healthy and functional food.

3.2.2 Cereals

Globally, cereals are regarded as a major source of protein, carbohydrates, vitamins, minerals, and fiber, particularly indigestible carbohydrates, or prebiotics, which exhibit numerous beneficial functional effects and serve as a food and energy source to the gut microbes like *Lactobacilli* and *Bifidobacteria*. These indigestible carbohydrates, or prebiotics in cereals, include water-soluble fiber (e.g., β-glucan and arabinoxylan), oligosaccharides (galactooligosaccharides [GOS] and FOS), and resistant starch. Interestingly, cereals have now become one of the preferred alternatives for the non-dairy industry as a probiotic carrier material owing to the aforementioned benefits. Among various cereals, oats, and barley have the highest content of β-glucan. Other cereals used for the development of non-dairy fermented cereal-based synbiotic products include maize, sorghum, millets, wheat, cornmeal, and rye.

Another benefit associated with fermented cereals is that fermentation induces the release of different organic acids and microbial enzymes, which in turn increases the availability of various minerals such as iron, phosphorus, and calcium. Several traditional cereal-based drinks are available in the market, which include Bushera, Boza, Pozol, Togwa, and Mahewu, made via the fermentation of cereals using a variety of yeasts and LAB as well as *Candida tropicalis*, *Geotrichum candidum*, *Saccharomyces cerevisiae*, and *G. penicillatum*. Moreover, recent studies have also explored the use of a combination of different cereals and probiotics to produce healthy synbiotic beverages in the non-dairy food industry.

3.2.3 Legumes and Pulses

Legumes are one of the staple diet constituents across the world. The most common and widely used legumes include kidney beans, mung beans, chickpeas, lentils, cowpeas, lupins, and soybeans. Among all the food groups, legumes and pulses are the richest sources of carbohydrates (30–60%), protein (19–36%), and dietary fiber (9–25%), along with amino acids like lysine, leucine, and arginine. The carbohydrates majorly include monosaccharides, oligosaccharides, polysaccharides, and starch. Legumes are also rich in minerals and vitamins and contain bioactive constituents like polyphenols that exhibit therapeutic potential due to their antioxidant activity. Some of these bioactive compounds in legumes include antinutritional factors (ANFs) like the RFOs, protease inhibitors, phytates, and saponins. These ANFs get fermented by the gut microbiota and produce hydrogen, carbon dioxide, and methane, causing flatulence, bloating, and discomfort. Nonetheless, these ANFs also serve as a component of dietary fiber and exhibit prebiotic properties by promoting the growth of beneficial bacteria, i.e., *Lactobacilli* and *Bifidobacteria*, in the gut. Hence, the fermentation of legumes using probiotics can be used to develop non-dairy synbiotic foods. Presently, fermented legume flour is widely used for developing bakery products (e.g., biscuits and bread) and alternative dairy products to enhance their technological properties and nutritional composition.

In the non-dairy synbiotic industry, the majority of research has been done using soybean as the main ingredient. Even at the commercial level, consumers, especially lactose-intolerant and vegan individuals, prefer soybean-based drinks with probiotics. Nonetheless, the recent advancement in the industry has also employed the use of different legumes like lupin, chickpea beans, peas, and lentils, as the potential ingredients for non-diary synbiotic alternatives for making healthy beverages, yogurts, ice cream, and other products.

3.2.4 Meat

Fermented meat products are regarded as the ideal matrix for probiotics among other processed meat products, primarily because of the absence of heat, and the sausage structure protects probiotics in the human GI tract to exhibit desirable health benefits. Among the vast fermentation ecosystem employed in the fermentation of meat, the most important ones are lactic acid bacteria and *Staphylococcus*, which are used as starter cultures. The concept of replacing traditional meat starter inoculums with probiotics has been considered a smart alternative method to develop healthy meat products. Moreover, probiotic microorganisms adapt easily to fermented sausages with high pH and salts like sodium chloride, sodium nitrate, and nitrite. Additionally, these fermented meat products protect the probiotics from the extreme gut environment (low pH and bile salts) and stimulate probiotic growth in the presence of prebiotic compounds. A lot of probiotic-based meat products are available in the market, which include salami, fuet, sturgeon-fermented sausage, and mutton-fermented sausage.

Furthermore, the addition of probiotic microorganisms to fermented meat products impacts the sensorial properties of the final product by altering pH and a_w. Hence, an appropriate quantity of probiotic microorganisms should be added to the meat products to maintain the final taste of the product. Moreover, the product should also meet the required counts of probiotics as per the recommended guidelines. In this regard, techniques such as immobilization, lyophilization, microencapsulation, and spray drying are used. They prevent the free cells from tolerating the harsh processing conditions as well as internal gut environment in order to maintain the required viability in the final synbiotic product.

3.2.5 Food Processing Waste

A major portion of the food and agriculture industry generates waste, which has now been used as a rich source of numerous secondary metabolites, including flavonols, anthocyanins, phenolic acids, etc. Various studies have shown the potential of these extracted metabolites as a new matrix for the delivery of probiotics and, thereby, the development of a novel synbiotic system. These prebiotic-rich waste products that have been explored in recent times include sugarcane bagasse waste, mucilage from faba bean and chickpea, and fruit peels and seeds that have been used to extract metabolites like xylo-oligosaccharides and other bioactive compounds and to develop healthy synbiotic foods and beverages.

3.3 The Prebiotic Activity of Non-Dairy Products

Different dietary substances act as a source of prebiotics in different food groups. However, in general, dietary fibers are the most considered potential prebiotics in the food industry. The basic property of a prebiotic substance is that it is not easily digested by humans and serves as a source of food energy to the gut microbiota. Prebiotics cause the growth of different indigenous gut bacteria and help in modifying the gut microbiota and changing the composition of the overall gut environment like the pH, temperature, and concentration. The prebiotics are also known to enhance the growth and overall characteristics of probiotics. Such a combination system is called synbiotics, and products employing this combination are referred to as synbiotic products. However, it must be noted that although all prebiotics are majorly fiber, but all fibers are not prebiotics.

Prebiotics are naturally found in non-dairy products like vegetables, fruits, cereals, pulses, and their wastes. Fruits, vegetables, cereals, millets, and legumes are some sources of natural prebiotics like starch, cellulose, hemicellulose, pectin, inulin, raffinose, verbascose, and stachyose, while artificially produced prebiotics include FOS, malto-oligosaccharides, cyclodextrins, and lacto-saccharose. Prebiotics by serving as food to probiotics and other gut microbes induce fermentation of prebiotics, which in turn releases SCFAs like acetic, butyric, and propionic acid. These SCFAs are taken up by the gut epithelial cells, thus regulating cell proliferation and differentiation along with other functions like mucus secretion and providing barrier integrity. Resistant starch is a prebiotic that produces a high level of SCFAs and is mostly used as an encapsulating agent in the food industry. Hence, plant-based food groups and products made from them can be employed as a smart and sustainable prebiotic source for the development of non-dairy synbiotic products with acceptable growth of probiotic microorganisms.

3.4 Processing and Formulation of Synbiotic Non-Dairy Products

The process flow chart for the formulation and processing of non-dairy synbiotic products is shown in Figure 3.2.

- Probiotic selection: The first step is the selection of the probiotic and its most suitable strain based on the application and the type of matrix used. The probiotic microorganism selected must be safe and able to survive in the harsh GI surroundings of the human gut.
- Prebiotic selection: The second step involves the selection of the potential non-dairy matrix (fruit/vegetable/cereal/legume/waste), which contains sufficient nutrients and prebiotic efficacy for the growth of the selected probiotic. The selected prebiotic must also exert beneficial properties and enhance gut health.
- Extraction (if required): The next step involves the extraction of required prebiotics or dietary fibers if required, such as in the case of food and agro-waste.
- Mixing: Followed by the selection and extraction, the next step involves mixing the prebiotic component with the probiotic strains. It can be achieved in two ways:

  1. With fermentation: By providing the two components with adequate conditions.
  2. Without fermentation: By encapsulating the probiotic in powder form by spray drying or freeze-drying and adding it to the non-dairy matrix with prebiotic efficacy.

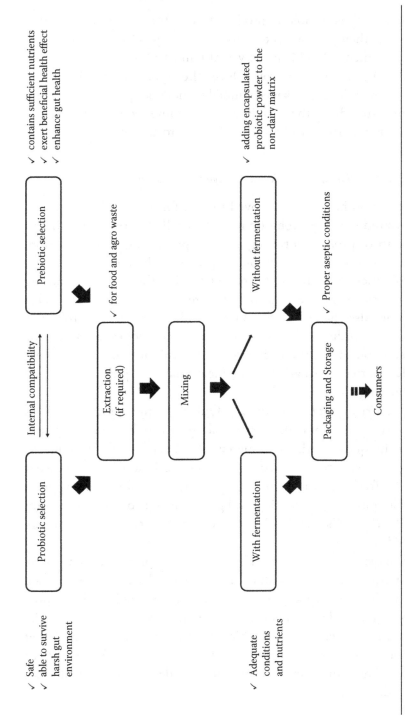

Figure 3.2 Process flow chart for the formulation and processing of non-dairy synbiotic food products.

The fortified synbiotic product from either of the two ways must then be subjected to a viability test as per the recommended guidelines by FAO and FSSAI.

• Packaging and storage: Once the synbiotic product is obtained, it is packaged suitably under aseptic conditions and stored at the required temperature conditions. The product is then ready for distribution to the consumers.

3.4.1 Encapsulation of Probiotics in the Synbiotic System

Encapsulation has been employed for the formation of a protective layer around the core probiotic microbial cells to prevent any injury and death of probiotics throughout the processing. Encapsulation, by acting as a protective membrane, preserves the metabolic activity of probiotics and prevents reduction in the viability during the processing as well as till storage of such products. This technique also helps in the uniform distribution of probiotics within the product. There are different techniques used for the encapsulation of probiotics in non-dairy synbiotic products. These include microencapsulation, nanoencapsulation, spray drying, and freeze drying.

Freeze drying and spray drying: Quick drying methods such as freeze drying and spray drying are the most widely used encapsulation techniques at the commercial level in the food industry. The final product developed using these techniques is in powder form. In freeze drying, a mixture of active probiotics along with desirable nutrients, prebiotics, and gelatin is mixed and poured into a sterile Petri dish and subjected to freeze drying in the range of 45–65°C.

In the case of spray drying, a mixture of the prebiotic-rich matrix along with probiotic inoculum in the desired percentage is used along with an additive gum. This mixture is subjected to spray drying with an inlet temperature in the range of 100–130°C (or the maximum temperature tolerable by the probiotic), an outlet temperature of 60–80°C, and a spray atomizer with concurrent airflow. The spray-dried synbiotic powder is then collected, packaged, and stored.

The advantage associated with encapsulation using these drying methods is the maintenance of the viable probiotic cell at the end of processing and shelf life. Moreover, the developed powder product is ready to use, convenient, and reduces transportation and storage costs. Hence, they are considered the best mode for the delivery of food products.

Microencapsulation: Microcapsules are generally developed using extrusion technology, wherein the extracted prebiotic is mixed in sodium alginate solution, followed by the addition of a lyophilized probiotic culture. The solution is mixed carefully using the stirrer to avoid any kind of cell distortion. This mixture is then added dropwise using a syringe into a slowly revolving solution of $CaCl_2$, which leads to the formation of small beads of sodium alginate mixture. The beads are then filtered and dipped into another $CaCl_2$ solution of different concentrations to ensure the hardening of the beads. After some time, the beads are separated and subjected to lyophilization.

Nanoencapsulation: Among all, nanoencapsulation is one of the new emerging technologies for the making of synbiotic products. Nanoparticles are known to improve the absorption of substances in the GI tract of the human body. So, prebiotics that have a limited absorption ability or poor bio-accessibility can be nano-encapsulated to increase the survivability of probiotic microbes even in unfavorable gut conditions. Nonetheless, other aspects associated with this new technology (e.g., toxicity) must also be explored while considering it as an option for the delivery of probiotics in food systems.

3.5 Health Benefits

Probiotics work as a gut health regulator, while prebiotics serve as an energy source for probiotics as well as humans. Prebiotic-rich fruits, vegetables, cereals, legumes, and millets are also good sources of proteins, minerals, fibers, vitamins, polyphenols, and antioxidants. In this context, a synbiotic non-dairy food item exhibits several health benefits as a result of synergy between the two components. In general, when the probiotics thrive and colonize the gut epithelium cells, they not just cause competitive exclusion of the pathogens but also stimulate the immune system, exhibit antimicrobial and anti-oxidant activity, and even have added properties by means of cell signaling. Apart from this, various studies have also proven certain specific benefits of synbiotics toward conditions like asthma, cancer, immune system, and mental health (Dahiya, Nigam, 2022).

3.5.1 Synbiotics for Reducing Asthma

The consumption of synbiotics made using the combination of *B. breve* (probiotic) with GOS and FOS (prebiotic) increased the peak expiratory flow in asthma patients. The treatment also reduced mediators responsible for causing inflammatory effects in the respiratory tract when exposed to allergens like IL-4, IL-5, and IL-13. Hence, the concept of synbiotic and synbiotic-based food products helps reduce the risk of allergic effects in asthmatic patients.

3.5.2 Synbiotics for Enhancing the Immune System

In general, both probiotic and prebiotic components are responsible for enhancing the immune system by releasing certain antibodies, which in turn are responsible for the regulation and improvement of the individual's immunity. Probiotics modulate innate immunity by exhibiting antiviral properties, wherein they increase the cytotoxic potential of natural killer cells as well as the phagocytosis property of macrophages. The probiotic component also stimulates the production of IgA, which binds antigens, thereby allowing modulation of the adaptive immune response. Prebiotics, on the other hand, improve the intestinal barrier and gut immune system via the production of SCFAs. Various studies on synbiotic-based food products for increasing immunity and allowing colonization of probiotic bacteria in the gut have been discussed and proven their ability successfully.

3.5.3 Synbiotics for Reducing the Risk of Cancer

Numerous *in vitro* and *in vivo* studies have indicated that probiotics like lactobacillus and bifidobacteria and prebiotics like oligofructose and inulin exert anti-neoplastic effects. Moreover, the combinations of probiotics and prebiotics in the form of a synbiotics system show greater efficacy toward the reduction of risk of, particularly, colorectal cancer (CRC), wherein the additive mechanism of both the components allows an increased immune response, decreased colonic inflammation, production of anti-tumorigenic compounds, and reduction of carcinogenic compounds. The use of probiotics and prebiotics has important implications in the field of CRC, as their consumption may be beneficial in inhibiting or preventing the onset of cancer and also in the treatment of existing tumors.

3.5.4 Synbiotics for Trauma Patients

Recent studies show the potential of synbiotic products with probiotics, namely, *L. mesenteroides, L. paracasei* ssp. *paracasei,* and *L. plantarum,* and prebiotics such as inulin, oat bran, pectin, and resistant starch in the treatment of acute illness of severely ill patients. Synbiotics maintain the gut flora and environment and reduce the chances of septic complications in patients. The intake of synbiotics increases the levels of beneficial bacteria such as *Bifidobacterium* and *Lactobacillus,* which in turn induces the production of SCFAs in the gut. These environmental changes can help to maintain the gut flora, enhance systemic immune function, and decrease the incidence of septic complications such as enteritis, pneumonia, and bacteremia. Hence, synbiotic non-dairy food items hold the potential to be explored in depth to help in reducing the risk and even treatment and prevention of different diseases and maintaining the overall well-being of consumers.

3.6 Challenges and Future Scope

The limitations and drawbacks shown by the dairy sector resulted in faster development and acceptability of non-dairy–based synbiotic food products inherently supporting the probiotic cultures. However, there are several challenges faced by the non-dairy industry to maintain its position in the food sector. Moreover, the addition of the concept of synbiotics also creates a part of the challenge to be addressed in the functional food industry.

Selection of the appropriate food matrix: The selection of an appropriate food matrix is crucial as the growth and viability of probiotics depend on it throughout the processing till the storage. Moreover, in the case of non-dairy matrices, the availability of prebiotic(s) in a good concentration is equally essential to fulfill the purpose of the synbiotic concept. Yet again, one should keep in mind that the prebiotic content of the non-dairy matrix should not be compromised with the physicochemical as well as the nutritional properties of the developed product. Hence, the selection of a matrix that has good nutritive value, acceptable acidity, and is rich in desirable prebiotics creates a challenge in the initial steps.

Stability and survivability of the probiotic strains: Along with the selection of an appropriate food matrix, the most suitable probiotic, as per the requirements of the food matrix, must also be considered. Since numerous factors influence the stability and survivability of the probiotic, like the water activity and pH of the prepared synbiotic product, the probiotic strain should be selected such that it remains stable in the product as such and during the processing. The probiotics must be able to tolerate the acidic environment of the product as well as the human gut. Thus, these conditions need maximum attention while selecting a probiotic strain; otherwise, the product might not meet the criteria of being called a probiotic or synbiotic product.

Shelf life: Another challenging factor that must be addressed while developing a synbiotic product is the attainment of the required viable count by the end of the shelf life; otherwise, there is no point in calling the product 'synbiotic.' In general, to enhance and maintain the shelf life of a product as well as probiotic viability, it is generally stored at lower temperatures, such as 4–5°C. Though non-dairy probiotic products generally require storage at ambient conditions, there is a need for the development and implication of new techniques for increasing the survivability of probiotics in these matrices as well as the shelf life of synbiotic products. Moreover, innovative packaging materials can also be used for the same purpose.

Acceptance: Once an appropriate combination of food matrix and probiotics is achieved, there comes the biggest challenge of the overall acceptance and sensory quality of the developed non-dairy synbiotic food items (Cosme et al., 2022). The interactions of different food ingredients along with probiotic strains are responsible for the production of different flavor compounds and changes in textures, color, and aroma of the final product. These changes in the sensory properties might differ from one individual to another. Moreover, the production of alcohol and gases as a part of fermentation in synbiotic products also affects their sensory properties. Therefore, there is a need for the development of synbiotic products that can be easily accepted by all age groups of society.

In conclusion, developing non-dairy probiotic food products is indeed a viable need of society in the present scenario. Despite these challenges and complexities at the technological level, many companies and startups in the food industry are coming up with innovative non-dairy synbiotic and probiotic-based food products that are receiving exceptional acceptance from the consumers, though the number remains less. Nevertheless, these products hold the potential to become successful products in the functional food industry, which requires great research to be undertaken at the same time. Moreover, there are so many unexplored food items and concepts within individual food groups that can be discussed in the future. For example, among legumes and pulses, only soybean has been researched extensively though there are several other prebiotic-rich pulses that can be combined with beneficial probiotics to develop non-dairy legume-based food products. Similarly, extraction of prebiotics from agricultural and processing waste can be explored to make innovative food items, which in turn will help in the reduction of food wastage in the environment.

3.7 Summary

Consumers who dislike dairy products, have lactose intolerance or have allergies to milk proteins, or prefer vegan foods may benefit a lot from the use of non-dairy–based products with additional health benefits conferred by the synbiotic system added to them.

Followed by dairy, the most common matrix for the delivery of probiotics belongs to the food groups of fruits, vegetables, cereals, millets, and pulses, along with meat and food and agro-waste as potential non-dairy products.

These food items also provide a source of prebiotics, hence completing the concept of synbiotics and exhibiting added nutrition to the product.

Non-dairy synbiotic foods and beverages have a longer shelf life, improved nutritional composition, and superior sensory quality, and are easily acceptable by the lactose-intolerant population of the world.

In the formulation of such non-dairy beverages, the overall effect can be increased by implementing techniques such as micro-encapsulation, nanoencapsulation, freezing drying, and spray drying.

Due to the numerous health benefits exerted by the individual components as well as the synergism between them, synbiotic non-dairy products can be considered an essential component in the treatment, prevention, and reduction of diseases like CRC, allergies, asthma, and overall development of gut health in trauma patients.

3.8 Multiple Choice Questions

1. Which among the following are the major food groups contributing to non-dairy synbiotic products?
   a. Fruits and vegetables
   b. Cereals and pulses
   c. Meat and food processing waste
   d. All of the above
2. Which food group is the second most widely used food group for the development of synbiotic food products after dairy?
   a. Fruits and vegetables
   b. Cereals and pulses
   c. Meat and food processing waste
   d. All of the above
3. Which among the following is the widely used technique (at commercial scale) for maintaining the efficacy of probiotics in non-dairy fermented beverages?
   a. Freeze drying
   b. Nanotechnology
   c. Spray drying
   d. All of the above
4. The presence of ____ in synbiotic products helps vegan people to obtain fibers that are generally obtained from meats.
   a. proteins
   b. dietary fibers

   c. short-chain fatty acids

   d. flavor compounds

5. The reason(s) for the shift from dairy to non-dairy synbiotics in the food industry include:

   a. Lactose intolerance

   b. Allergy to milk and dairy products

   c. Veganism

   d. All of the above

6. The major source of prebiotics obtained from legumes is/are:

   a. Oligosaccharides

   b. Glucosidase

   c. Lactose

   d. Monosaccharides

7. _____ and galactooligosaccharides are the two important groups of prebiotics with beneficial effects on human health.

   a. Fructooligosaccharides

   b. Raffinose oligosaccharides

   c. Inulin

   d. Resistant starch

8. Fruits such as bananas contain _____ as prebiotic(s).

   a. resistant starch

   b. cellulose and hemicelluloses

   c. lignin

   d. All of the above

9. The most widely used probiotic strain for the development of synbiotic non-dairy products using fruits and vegetables is _____.

   a. *Lactobacillus plantarum*

   b. *Streptococcus thermophilus*

   c. *Bacillus coagulans*

   d. *B. animalis*

10. In the formulation of synbiotic non-dairy products via microencapsulation, the prebiotics are combined with _____ solution, and then lyophilized probiotics are added.

   a. sodium bicarbonate

   b. sodium alginate

   c. calcium chloride

   d. calcium acetate

3.9 Short Answer Type Questions

Q1. Write a short note on each topic: the potential of the non-dairy–based matrix for the development of synbiotic foods; the future scope of food processing waste as a potential source of prebiotics for making synbiotic food products.

Q2. List the main reasons causing a shift in interest from dairy to non-dairy–based synbiotics food products.

Q3. List the different techniques used for the formulation of non-dairy synbiotic food products.

Q4. Highlight the potential of meat products to serve as a prebiotic source for a non-dairy synbiotic food product.

3.10 Descriptive Questions

Q1. Explain in brief 'fruits and vegetables as a non-dairy synbiotic matrix.

Q2. How do cereals and pulses exhibit themselves as a potential prebiotic source in comparison to dairy products?

Q3. Explain in brief 'food processing waste' as a non-dairy synbiotic matrix.

Q4. Discuss the prebiotic activity of non-dairy products.

Q5. Describe the processing and formulation of synbiotic non-dairy products in detail.

3.11 Answers for MCQs

Q1	Q2	Q3	Q4	Q5	Q6	Q7	Q8	Q9	Q10
d	a	c	c	d	a	a	d	a	b

References

Chaturvedi, S., & Chakraborty, S. (2021). Review on potential non-dairy synbiotic beverages: A preliminary approach using legumes. *International Journal of Food Science & Technology*, 56(5), 2068–2077. 10.1111/ijfs.14779

Kumar, D., Lal, M. K., Dutt, S., Raigond, P., Changan, S. S., Tiwari, R. K., Chourasia, K. N., Mangal, V., & Singh, B. (2022). Functional fermented probiotics, prebiotics, and synbiotics from non-dairy products: A perspective from nutraceutical. *Molecular Nutrition & Food Research*, *66*(14), 2101059. 10.1002/mnfr.202101059

Mishra, A., Chakravarty, I., & Mandavgane, S. (2021). Current trends in non-dairy based synbiotics. *Critical Reviews in Biotechnology*, *41*(6), 935–952. 10.1080/07388551.2021.1898329

Suggested Readings

Dahiya, D., & Nigam, P. S. (2022). Nutrition and health through the use of probiotic strains in fermentation to produce non-dairy functional beverage products supporting gut microbiota. *Foods*, *11*, 2760. 10.3390/foods11182760

Cosme, F., Inês, A., & Vilela, A. (2022). Consumer's acceptability and health consciousness of probiotic and prebiotic of non-dairy products. *Food Research International*, *151*, 110842. 10.1016/j.foodres.2021.110842

4

Synbiotics in Cereal Industry

4.1 Cereal and Cereal Products for the Development of Synbiotic Products

Among the different food groups, cereals are rich sources of carbohydrates, along with minerals, vitamins, protein, and fiber. Cereals comprise numerous prebiotic constituents that benefit humans by selectively stimulating the growth of good residential bacteria in the gut by acting as an energy source to them. Apart from this, cereals are recognized as the most suitable substrates in the food industry for fermentation. For this reason, cereals and cereal-based products are used as delivery matrices for various food formulations. Additionally, prebiotic substances such as water-soluble fibers, oligosaccharides, and resistant starch have shown their potential to enhance the growth of probiotic microbes along with the gut microflora. In this way, the formulation of novel, innovative, and healthy 'synbiotic' products can be carried out. Such cereal-based foods fermented using probiotics not only improve consumer's health but also serve as a healthy non-dairy food option for lactose-intolerant and vegan populations. Even the literature shows extensive research done on exploring the prebiotic ability of various cereals. The major probiotics belong to the family of Lactobacillus, namely, *L. plantarum*, *L. acidophilus*, and *Bifidobacterium* spp. Similarly, the main cereals used include wheat, maize, oats, rye, millet, barley, and rice.

Numerous researchers have explored the potential of these prebiotic-rich cereal substrates for specific probiotic microorganisms to develop a synbiotic system to deliver maximum health benefits to the host (Figure 4.1). Moreover, the combination of different food groups, for example, combining vegetables or dairy with cereals, to make different synbiotic food products has also been studied. Such combinations are beneficial, being the chief source of nutrients like polyphenolic compounds, ascorbic acid, and carbohydrates, which may lack in the

 DOI: 10.1201/9781003304104-4

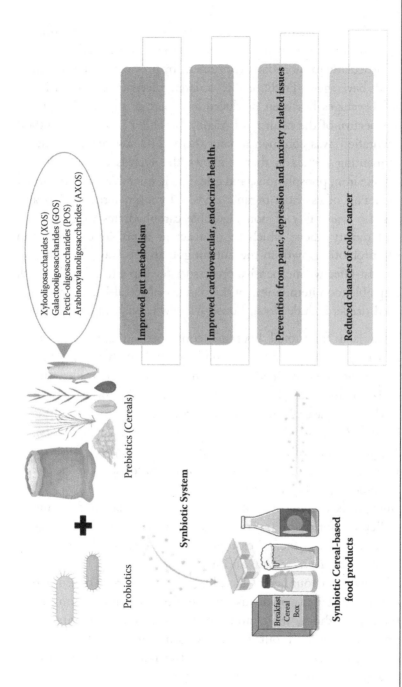

Figure 4.1 Health benefits of synbiotic cereal-based food products in humans.

counterpart. Thus, it serves as an improved substrate for fermentation by gut microbes as well as the probiotic lactic acid bacteria.

Cereals are consumed worldwide as a daily staple food contributing a major portion of the human diet. They are also considered a suitable choice, particularly for non-dairy consumers, by providing ready-to-drink fermented beverages. Traditionally, a large diversity of cereal-based probiotic beverages has been produced as a part of the regular human diet. Irrespective of the country of origin, the oldest processing method of 'fermentation' was common among them. The advantage associated with fermenting cereals is that it allows the increased availability of minerals like iron, phosphorous, and calcium via microbial enzymes or by the production of essential organic acids. Moreover, fermentation also increases the availability of non-digestible carbohydrates (oligosaccharides) and dietary fibers, which in turn act as a prebiotic source and stimulate probiotic growth in the human gut. In this way, fermentation acts as a cost-effective inexpensive process to develop a healthy, flavorful, and acceptable food item. Some of the traditional cereal-based probiotic-fermented drinks include Boza, Pozol, Mahewu, Bushera, and Togwa.

- Boza is a traditional beverage consumed in countries like Albania, Bulgaria, Romania, and Turkey. Usually served cold, this beverage is obtained by fermenting wheat, rye, millet, and maize, combined with sugar. Boza utilizes a wide variety of LAB and yeast cultures for the fermentation process. These include *L. acidophilus, L. brevis, L. coprophilus, L. fermentum, L. plantarum, Leuconostoc mesenteroides, Leuconostoc reffinolactis, C. glabrata, C. tropicalis, Saccharomyces cerevisiae, Geotrichum candidum,* and *G. penicillatum.*
- Bushera is the traditional beverage of Uganda, made using millet or sorghum (germinated) flour and probiotics like *L. brevis* and others belonging to the genera *Lactococcus, Leuconostoc, Enterococcus,* and *Streptococcus.*
- Mahewu is a sour cereal-based probiotic beverage consumed in South Africa, made using a multi-grain mix of millet, maize, sorghum, malt, and wheat flour. The fermentation process is carried out by the existing microbiota of the cereals, especially malt. However, the main microorganism includes *Lactococcus lactis* subsp. *lactis.*

Table 4.1 Summary of Research Done on the Formulation of Synbiotic Cereal-Based Food Product

PRODUCT	PROBIOTIC	PREBIOTIC	KEY CONCLUSION	REFERENCES
Synbiotic-fermented finger millet-based yogurt-like beverage	Co-culturing *Weissella confusa* 2LABPT05 and *Lactiplantibacillus plantarum* 299 v	Exopolysaccharide from African finger millet	• Co-culturing allowed the best fermentative performance with probiotic count above 10^8 CFU/mL. • Developed product showed improved nutritional and physicochemical profiles, increased protein digestibility, and desirable viscosity.	Vila-Real et al. (2022)
Synbiotic beverage based on millet, rye, and alfalfa sprouts	*Lacticaseibacillus casei* and *L. plantarum*	Inulin and oligofructose	• Synbiotic beverage contained >10^6 CFU/mL probiotics throughout the storage period. • Beverages showed acceptable sensory characteristics.	Mohammadi et al. (2021)
Synbiotic yoghurt using brittle of walnut and cereal flakes (barley, wheat, rye, oats, and buckwheat)	*L. bulgaricus*, *L. casei*, and *Streptococcus thermophilus*	Prebiotics from the selected cereals and walnut brittle	• Developed formulation showed enhanced nutritional and biological values. • The finished product was also desirable from organoleptic, physicochemical, and microbiological point of view.	Gorlov et al. (2019)
Synbiotic rice-based yogurt-like product	*Bifidobacterium* and *L. acidophilus*	Inulin	• Developed product had 2.0% protein and 1.1% fat; good sensory attributes and sensory score of 4.3 out of 5. • Probiotic's count remained above 10^6 CFU/mL after eight weeks of storage.	Wang et al. (2021)
Synbiotic cereal-based vegan dessert using millet-oat flour mixture	*L. plantarum* 020	Inulin	• Probiotic's count was in desirable range even after 21 days of refrigerated storage. • Sensory assessment indicated that the formulation with 3% *w/v* inulin showed highest overall quality.	Szydłowska et al. (2021)

- Togwa is another most-consumed probiotic product from Japan and China. It is made by fermenting finger millet and maize flour with probiotic bacteria like *L. plantarum* and *Streptococcus*.
- Pozol is another traditional refreshing, acidic, and non-alcoholic fermented beverage made using nixtamalized maize dough. It is a common food option in Southeastern Mexico, which utilizes the probiotic *Lactococcus lactis*.
- Ogi, also known as Akamu, is a fermented cereal product, popular in Nigeria and west Africa, typically made from maize, sorghum, and millet and fermented using a wide variety of probiotics like *L. acidophilus, L. fermentum, L. plantarum, L. brevis, S. cerevisiae, Candida krusei, C. tropicalis, Rhodotorula graminis, G. candidum,* and *G. fermentum*.

Several studies have been carried out to develop synbiotic cereal-based beverages and other food products and to determine the suitability of different cereals to improve the growth of probiotics and maintain their viability. Table 4.1 summarizes the recent work in this regard.

4.1.1 Oat-Based Products

Oat (*Avena sativa*) is a rich source of dietary fiber (both insoluble and soluble) along with minerals like selenium, phosphorus, thiamine, magnesium, zinc, and other phytochemicals significant for human health. Among the non-digestible dietary fibers, oat β-glucan has shown the most beneficial effects toward hypertension, obesity, and dyslipidaemia. It also enhances the immune response toward bacterial attack and infection and is even used in the treatment of cancer and its prevention. In humans, oat β-glucan acts as a prebiotic in the digestive system and carries out selective fermentation via butyrate-producing microorganisms.

The literature identifies the application of exopolysaccharides (EPS) producing *Pediococcus parvulus* 2.6 in fermented oat-based products to improve the overall viscosity, texture, and mouthfeel of the product. Even trials on humans have shown the combination of this synbiotic system in decreased cholesterol levels and increased *Bifidobacterium* spp. count in the gut. Studies have also concluded the potential of EPS via *P. parvulus* 2.6 as a prebiotic substrate with its ability to enhance the growth of probiotic LAB strains in the human gut. Probiotic oat-based beverages fermented using different *Lactobacillus* strains have also shown a high antioxidant activity and enhanced total polyphenol content in comparison to other non-fermented beverages. Several studies tried to develop synbiotic oat-based beverages using probiotic cultures like *L. plantarum* B28, *L. plantarum* LP09, *L. curvatus* P99, and oat prebiotic β-glucan. The results of these investigations have proven the synbiotic potential of such combinations. The fermentation tends to increase the β-glucan content which is associated with a decrease in the soluble and insoluble fiber content. This decrease is an indication of uptake of the prebiotic components by the probiotic microbes. Moreover, from sensorial point of view, the aroma and flavor of the developed synbiotic products is superior from the non-fermented control samples with greater consumer acceptability.

4.1.2 Malt-Based Products

Malt or the germinated cereals has been known as a suitable medium for probiotics to grow, thrive, and maintain their count in the developed product throughout the storage life. The higher viability of the probiotics in malt-based synbiotic products is attributed to the nutritional changes undergone by the cereal via germination that increases the content of sugars like glucose, sucrose, maltose, fructose, maltotriose, and maltotetraose in the malt, which acts as a prebiotic substrate for them. Studies have also shown that malt-based matrices tend to favor the growth of *Bifidobacterium* spp. This is because of the presence of sugars like maltotriose and glucose that are more preferably utilized by probiotics from *Bifidobacterium* family including *B. adolescentis*, *B. breve*, and *B. longum*.

4.1.3 Wheat-Based Products

Emmer wheat flour (*Triticum dicoccon*) is made using hulled wheat and is an ancient wheat type, an ancestor of the modern durum wheat. This flour has been successfully employed in the production of pro/synbiotic emmer beverages owing to its high resistant starch content, which acts as a prebiotic component. The probiotic-emmer beverages are made using emmer flour, emmer gelatinized flour, and emmer malt. These beverages are fermented using probiotic strains like *L. plantarum* 6E and *L. rhamnosus* SP1. Probiotic beverages made using such synbiotic combinations have shown desirable viability and sensory properties throughout the shelf life of the product. Moreover, the recent addition of *S. cerevisiae* var *boulardii* as a potential probiotic for the development of alcoholic wheat beverage, 'wheat beer', has shown desirable results. The probiotic yeast holds the ability to exhibit equivalent metabolic efficacy and growth on the wort as that of the usual brewer's yeast. Moreover, this strain is even resistant to the processing conditions and GI transit. Wheat in combination with other cereals like malt and barley extracts can also be used as a potential prebiotic substrate to be utilized by probiotics like *L. plantarum, L. acidophilus,* and *L. reuteri.* Such a combination allows the consumption of maximum sugars and enhanced viability of probiotics in the fermented beverages. Recent studies have also employed probiotics like *L. rhamnosus* GG which releases phenolic acids like trans-ferulic acid during the colonic fermentation of the whole wheat products (e.g., bread, cookies, and pasta). This in turn is responsible for the improved colon health in humans. Hence, the incorporation of appropriate probiotic microbes into wheat-based products hold the ability to serve as a delivery matrix and a prebiotic source with additional health benefits for health-conscious consumers.

4.1.4 Rice-Based Products

Rice (*Oryza sativa*, L.) is a popular base ingredient for numerous rice-based fermented drinks and foods, especially in the Asia-Pacific region. Rice beer is among such products with its popularity in the cultural societies of different parts of India. Mostly, probiotics are used for carrying out the fermentation in these traditional food items. For example, probiotics like *L. fermentum* KKL1 and *Bacillus velezensis* strain DU14 are used for fermentation of the traditional rice beers Haria and Apong, respectively. Additionally, the probiotics improve the buildup of minerals (Ca, Fe, Mg, Mn, and Na) and enhance the digestibility via enzymatic activity carried out by α-amylase, glucoamylase, etc. Moreover, the probiotic components also exhibit therapeutic abilities in terms of increased antioxidative properties. Similarly, fermented sour rice is another traditional food item consumed on a regular basis in India. Studies have confirmed the therapeutic and prophylactic abilities associated with the uptake of this food against different health issues. For the fermentation of this food item, the probiotic, *Weissella confusa* strain GCC 19R1 is predominantly used. Thus, these studies confirm the use of rice as an efficient matrix for probiotics like *L. fermentum* KKL1 and *L. plantarum* for their growth and existence. Furthermore, the synergistic effect of these probiotics on the prebiotic sugars in the rice-based products enhances their functional properties. Nonetheless, the need for *in vivo* analysis is essential to explore and validate the synbiotic potential.

4.1.5 Maize-Based Products

Maize (*Zea mays*, L.) is another popular cereal consumed globally. It has a nutrient-dense composition with about 72% starch, 10% proteins, remaining fiber, vitamin B complex, and essential minerals. In the beverage industry, too, maize is a widely used raw material to produce fermented and non-fermented drinks. Among the fermented ones, maize has proven itself as a potential matrix to develop probiotic beverages via yeast-lactic fermentation. For instance, a combination of potentially probiotic yeasts, namely, *S. cerevisiae* and *Pichia kluyveri*, together with commercial probiotic *L. paracasei*, has been used to develop a novel, functional fermented beverage. Interestingly, the developed beverage lacked any sweetness with no flavoring effect but showed desirable consumer acceptance. Apart from being a source of prebiotics and delivery matrix to the available probiotic microbes, the fermentation step can also be used for the detoxification purposes. Probiotics in the fermented food medium reduce toxins, which otherwise can negatively impact the functioning of immune system and affect nutrient absorption inducing liver cancer. In this regard, a study investigated the effect of probiotic addition in reduction of aflatoxins in Kwete beverage, which is a traditional African drink made using fermented maize. Thus, the drink produced by fermenting using probiotic *L. rhamnosus yoba* and *S. thermophilus* reported a significant reduction in the aflatoxin's levels. Such an approach of detoxification is of novel use owing to the wide consumption in countries like Africa. It can be noticed that most of the synbiotic grain-based products are traditional in nature; thus, the commercial availability as well as value addition is required to allow their acceptance globally.

Apart from this, the application of cereal blends for developing fermented food and drinks has also been studied lately, with maize being one of the major ingredients in such products. For example, a probiotic porridge from fermented maize has been prepared using barley and maize flour and probiotics like *L. rhamnosus GG*, *L. acidophilus LA5*, *L. acidophilus 1748*, or *L. reuteri*. Similarly, mixed beverages from maize and rice fermented via *L. pantarum* and *L. acidophilus* are some of the most explored matrices. The results of these studies reported good functional properties and acceptable sensory scores, indicating their high potential for the functional food market.

Thus, maize-based fermented beverages and food items have shown a potential to be utilized as a growth matrix for the probiotics, thereby incorporating synbiotic concept efficiently. However, future studies may focus on identifying the potential bioactive compounds in these food products, followed by their quantification and effect on the modulation of the intestinal microbiota. Such interventions could be of great interest to humans to understand the importance of maize beverages in the food industry.

4.1.6 Millet-Based Products

In comparison to some major cereals like rice and wheat, millets tend to have a higher nutritive value, being a rich source of proteins with essential amino acids like methionine. Millets also contain beneficial phytochemicals and micronutrients. In addition, millets hold the potential to exhibit themselves as a potential prebiotic source and allow the growth of major probiotic microorganisms.

Among the different millets, pearl millet is one such millet with a rich composition in proteins, soluble and insoluble dietary fibers, resistant starch, macro and micro minerals, and dietary antioxidants like ferulic acid and β-carotene. A traditional pearl millet porridge (kambu koozh) fermented using probiotic *Lactobacillus* strains including *L. fermentum* and *L. delbrueckii* has shown its ability to produce β-galactosidase and glutamate decarboxylase enzymes, which in turn have cholesterol-lowering effects in humans. In the same way, probiotic strains of *Lactobacillus* and *Pediococcus*, namely, *L. plantarum* subsp. *plantarum*,

L. sakei subsp. *sakei*, *L. pentosus*, *P. acidilactici*, and *P. pentosaceus*, isolated from a traditional alcoholic drink of South Korea, 'Omegisool', have shown desirable antioxidative properties. Similarly, other millets like foxtail millet (*Setaria italica*) and sorghum are being used in the functional food industry to formulate synbiotic drinks with the added advantages of both probiotic and prebiotic components. The literature also explores the possibility for developing synbiotic millet-based baked item. For example, recent studies have explored the application of sorghum flour and soybean flour as the main raw materials in the manufacture of synbiotic biscuits with freeze-dried *L. acidophilus* probiotic culture. This indicates the potential of millets to be utilized as a prebiotic matrix for the development of synbiotic functional food owing to the increasing demand of the health-conscious consumers in the society.

4.1.7 Fermented Pseudocereal Beverages

Among the cereal food group, the use of pseudocereal for the development of non-dairy pro/synbiotic beverages as a suitable delivery vehicle for bioactive compounds, probiotics, and prebiotics has been explored recently. Pseudocereals are considered a practical substrate for fermentation via probiotics because of their nutrient-dense composition. They are high-quality protein rich, gluten-free plants with minerals (Ca, Cu, Mg, Mn, Fe, and Zn) in greater quantities than in the basic cereals.

Among all, quinoa (*Chenopodium quinoa*, L.) is the most extensively consumed pseudocereal, which is known to contain all the essential amino acids (lysine, methionine, and threonine). It can also decrease the risk of type 2 diabetes and cardiovascular diseases. Furthermore, it is gluten-free, so its consumption is suitable for celiacs and people with gluten-allergy problems. Studies on the use of quinoa as a suitable food matrix for the development of synbiotic fermented beverages have enhanced its use in the functional food industry. The most common probiotics for its fermentation include *L. plantarum*, *L. casei*, and *Lactococcus lactis*. Such synbiotic beverages have a positive effect on the GI tract of the host as they allow the growth of beneficial microbes and production of SCFAs. Moreover, a significant decrease in the ammonia ion production and other related toxic substances is also associated with the uptake of such products. Nonetheless, there is a need for more long-term human clinical trials to demonstrate the probiotic efficacy of these synbiotic beverages.

Other studies have also explored the potential of other pseudocereals like amaranth (*Amaranthus*, L.), chia (*Salvia hispanica*, L.), and buckwheat (*Fagopyrum esculentum*, L.) as suitable matrices to make synbiotic pseudocereal-based food products and beverages. It has been depicted by the results of these studies that, either individually or in combination, these pseudocereals can be fermented using an appropriate probiotic culture along with any additional prebiotics. In this way, a synbiotic product with a desirable probiotic count, physiochemical and sensorial properties can be produced. Moreover, being gluten-free in nature, pseudocereal-based food products can also be used by the population that has conditions like gluten intolerance or related allergies. Such products are potential dairy substitutes with added health benefits exhibited by the synergistic mechanism between probiotics and prebiotics.

4.2 Summary

Cereals are considered a potential viable substrate for the development of pro/synbiotic-based food products because of the presence of the nutrients in them that are easily assimilated by probiotic microorganisms.

Cereals serve as an efficient transporter of lactic acid bacteria through the severe GI conditions.

Cereals have shown properties to stimulate the growth of single- as well as mixed-culture fermentations and can be used to formulate synbiotic cereal-based beverages with probiotic efficacy.

Various traditional probiotic-based cereal food products are among the widely consumed food commodity worldwide, making cereals a superior and most suitable probiotic matrix.

An efficient application of these cereals along with the scope for more innovation is required in the functional food industry to provide the consumers with more healthy and economical food products.

Among all, oats and rice are the most explored cereals for making synbiotic cereal-based products owing to their high prebiotic content. Additionally, studies have shown that beverage-based food products are most suitable for making synbiotic products because of their suitable matrix for both prebiotic extraction and probiotic proliferation.

4.3 Multiple Choice Questions

1. Typically the prebiotic components in cereals are _____.
   a. water-soluble fibers
   b. oligosaccharides
   c. resistant starches
   d. All the above

2. The major probiotics used for fermentation in cereals belong to the family of ___.
   a. *Lactobacillus*
   b. *Bifidobacterium*
   c. Both a and b
   d. None of the above

3. The main cereals used in the functional food industry include _____.
   a. wheat, maize, oat, and rye
   b. millet, barley, and rice
   c. pseudocereals
   d. All of these

4. Fermentation of cereals is beneficial as _____.
   a. it increases the availability of minerals via microbial enzymes
   b. it increases the availability of minerals via production of essential organic acids
   c. it increases the availability of dietary fibers and oligosaccharides
   d. All of the above

5. Boza is obtained by fermenting _____.
   a. wheat, soybean, and maize, combined with sugar
   b. wheat, rye, millet, and maize combined with sugar
   c. wheat, rye, millet, and maize combined without sugar
   d. wheat, soybean, and maize, combined without sugar

6. Traditional beverage made using millet or sorghum (germinated) flour is termed as _____.
   a. Boza
   b. Bushera
   c. Pozol
   d. Mahewu

7. The major prebiotic in oats is _____.
   a. β-glucan
   b. β-glycosidase
   c. Both a and b
   d. None of the above
8. Fermentation by probiotic reduces _____ in maize.
   a. aflatoxins
   b. probiotics
   c. SCFAs
   d. All of these
9. Most extensively consumed pseudocereal known to contain all essential amino acids is ____.
   a. maize
   b. sorghum
   c. quinoa
   d. wheat
10. The millet often used to produce synbiotic products is ____.
    a. pearl millet
    b. foxtail millet
    c. sorghum
    d. All of these

4.4 Short Answer Type Questions

Q1. Write a short note on 'cereal and cereal products for the development of synbiotic products.'
Q2. Mention the main probiotics and cereals used in the production of synbiotic cereal-based products.
Q3. Enlist the different traditional cereal-based pro/synbiotic products.
Q4. Discuss in short: Importance of fermentation in cereal-based products.
Q5. Enlist the advantages associated with synbiotic cereal-based products in the functional food industry.

4.5 Descriptive Questions

Q1. Enlist the main cereals contributing to the synbiotic cereal industry. Explain any two in detail.

Q2. With a process flow chart, discuss the critical control points while manufacturing a traditional probiotic cereal-based fermented beverages.

Q3. Discuss the manufacturing step of any traditional synbiotic cereal-based food product.

4.6 Answers for MCQs

Q1	Q2	Q3	Q4	Q5	Q6	Q7	Q8	Q9	Q10
d	c	d	d	b	b	a	a	c	d

References

Gorlov, I. F., Shishova, V. V., Slozhenkina, M. I., Serova, O. P., Zlobina, E. Y., Mosolova, N. I., & Barmina, T. N. (2019). Synbiotic yoghurt with walnut and cereal brittle added as a next-generation bioactive compound: Development and characteristics. *Food Science & Nutrition*, 7(2019), 2731–2739.

Mohammadi, M., Nouri, L., & Mortazavian, A. M. (2021). Development of a functional synbiotic beverage fortified with different cereal sprouts and prebiotics. *Journal of Food Science and Technology*, 58(11), 4185–4193. 10.1007/s13197-020-04887-4

Szydłowska, A., Siwińska, J., & Kołożyn-Krajewska, D. (2021). Cereal-based vegan desserts as container of potentially probiotic bacteria isolated from fermented plant-origin food. *CYTA – Journal of Food*, 19(1), 691–700. 10.1080/19476337.2021.1963320

Vila-Real, C., Pimenta-Martins, A., Mbugua, S., Hagrétou, S. L., Katina, K., Maina, N. H., Pinto, E., & Gomes, A. M. P. (2022). Novel synbiotic fermented finger millet-based yoghurt-like beverage: Nutritional, physicochemical, and sensory characterization. *Journal of Functional Foods*, 99(July). 10.1016/j.jff.2022.105324

Wang, C., Li, D., Wang, H., & Guo, M. (2021). Formulation and storage properties of symbiotic rice-based yogurt-like product using polymerized whey protein as a gelation agent. *CYTA – Journal of Food*, 19(1), 511–520. 10.1080/19476337.2021.1923573

5

SYNBIOTICS IN LEGUME INDUSTRY

5.1 Legumes and Legume-Based Products for the Development of Synbiotic Foods

Legumes have always been known for their rich carbohydrate, dietary fiber, and protein profiles. However, recent studies have also explored that legumes are also a potential and suitable medium for fermentation via beneficial microorganisms. Among the different food groups, dairy, fruits, vegetables, and cereals have been commonly consumed as probiotic products in the commercial market. Scientists have also proven the potential of these food groups as suitable probiotic carriers with the required prebiotic content to exhibit an efficient synbiotic system. Such products are favored over supplements for being a fresh and less processed commodity. However, when we are talking about functional foods, another major group, 'legumes and pulses,' still has not made it to the commercial scale yet. Though most of the conventional fermented foods were made using legumes, legumes are not explored to much of their extent in formulating probiotic food items. Legumes are rich sources of resistant starch and oligosaccharides or 'prebiotics' that can be employed to develop a non-dairy synbiotic food product. Alongside, fermentation with probiotics, the typical 'beany' flavor associated with legume-based products also reduces. Fermentation also enhances the shelf life of the product. Such a system also exhibits numerous beneficial effects to human beings, thus encouraging its acceptance in non-dairy products. The benefits include adequate colonization of probiotics in desirable limits during processing as well as storage, probiotic adherence to the gut lining, and modulation of essential immune-related functions. Moreover, the physical structure of legumes also provides a secure atmosphere for probiotics to tolerate and survive harsh gastric conditions. Additionally, such a system also provides a dietary option to lactose-intolerant people or those who have soy/milk/cholesterol-related allergies (Figure 5.1).

DOI: 10.1201/9781003304104-5

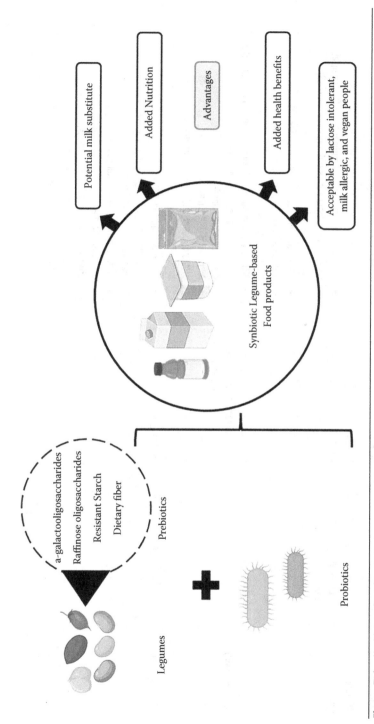

Figure 5.1 Health benefits of synbiotic legume-based food products in humans.

Table 5.1 Summary of Recent Studies Done on the Formulation of Synbiotic Legume-Based Food Products

SYNBIOTIC LEGUME-BASED FOODS	PROBIOTICS	PREBIOTICS	KEY TAKEAWAYS	REFERENCE
SOYBEAN				
Soymilk	*Lactobacillus acidophilus*, *Bifidobacterium lactis*,	Inulin	• Synbiotic soy milk has no significant effect on blood parameters but improves serum lead levels. • Study provided a simple and effective method of preparing synbiotic milk with potential chelation effects.	Riasatian et al. (2023)
Soymilk powder (SMP)	*L. plantarum* JlBYG12	Inulin, xylooligosaccharides, fructooligosaccharide, and isomalto-oligosaccharide (IMO)	• SMP resulted in increased bone density, improved bone metabolism, and enhanced bone remodeling. • Addition of 1.2% IMO increased the probiotic count and calcium enrichment by 8.15% and 94.53%.	Jia et al. (2023)
Soymilk yogurt	*LactoBacil* Plus (SCLBP) probiotic cultures	Fructooligosaccharide	• Antioxidant activity (75.28%) in synbiotic yogurt was significantly greater than control soy yogurt (52.98%). • In mice with hypercholesterolemia, synbiotic soy yogurts showed favorable effects in reducing blood cholesterol, low-density lipoprotein cholesterol, and lipid peroxidation in liver.	Sengupta et al. (2019)

(Continued)

Table 5.1 (*Continued*) Summary of Recent Studies Done on the Formulation of Synbiotic Legume-Based Food Products

SYNBIOTIC LEGUME-BASED FOODS	PROBIOTICS	PREBIOTICS	KEY TAKEAWAYS	REFERENCE
CHICKPEA				
Chickpea milk (CPM)	*Bacillus subtilis*	Resistant starch fibers	• Desirable colonization of CPM with wild-type cells of *B. subtilis* displayed enhanced survivability and resilience to environmental stress, such as heat and *in vitro* gastrointestinal (GI) treatments. • Biofilm formation on CPM fibers is an adaptation response of *B. subtilis* for strategic survival.	Rajasekharan et al. (2021)
KIDNEY BEAN				
Red kidney bean beverage (RKB)	*Lacticaseibacillus casei*	Raffinose, stachyose from kidney bean	• Optimized condition for formulating RKB involves extraction at 89 °C for 12.3 min using 0.01% $NaHCO_3$. • The developed beverage showed maximum prebiotic content, good yield, and total viable count, along with reduced antinutrients and acceptable sensory scores.	Chaturvedi and Chakraborty (2022b)
Kidney bean beverage powder	*Lacticaseibacillus casei*	Raffinose, stachyose from kidney bean	• Spray-dried gum acacia-coated powder showed maximum yield, better encapsulation efficiency (79.9–89.3%). as well as lower moisture content values (<5%).	Chaturvedi and Chakraborty (2022a)

(Continued)

Table 5.1 (*Continued*) Summary of Recent Studies Done on the Formulation of Synbiotic Legume-Based Food Products

SYNBIOTIC LEGUME-BASED FOODS	PROBIOTICS	PREBIOTICS	KEY TAKEAWAYS	REFERENCE
Kidney bean beverage yogurt	Vegan yogurt starter containing several strains of Lactobacteria, including *L. acidophilus, L. casei, L. delbrueckii subsp bulgaricus,* and *L. rhamnosus*	Raffinose, stachyose from kidney bean	• Probiotic viability was also >7 log CFU/mL in the case of GI simulation and under gastric acid and bile juices. • Yogurts showed stronger cellular antioxidant and anti-inflammatory activities because of the higher total and individual phenolic and peptide contents. • Peptide fractions of yogurts containing higher concentrations of γ-glutamyl-peptides showed stronger anti-inflammatory activity in Caco-2 cells.	Chen et al. (2019)
BAMBARA GROUND NUT				
Bambara groundnut beverage	*L. bulgaricus* and *L. plantarum*	Dietary fiber from Bambara groundnut	• Beverage had a shelf life of 2 and 14 days at accelerated storage 25 °C and under refrigeration (5 °C), respectively. • Probiotic viability remained >10^7 CFU/mL till the end of shelf life.	Murevanhema and Jideani (2020)

(Continued)

Table 5.1 (Continued) Summary of Recent Studies Done on the Formulation of Synbiotic Legume-Based Food Products

SYNBIOTIC LEGUME-BASED FOODS	PROBIOTICS	PREBIOTICS	KEY TAKEAWAYS	REFERENCE
Bambara groundnut yogurt	L. delbrueckii subspp. Bulgaricus, and Streptococcus salivarus subspp. thermophilus	Dietary fiber from Bambara groundnut	• The pH of yogurt decreased during storage along with simultaneous increase in titratable acidity although the sensory acceptability was higher. • Probiotics were in a desirable range till the end of storage while pathogens (Salmonella, Coliform, and E. coli) were absent.	Falade et al. (2015)
Bambara groundnut milk (BGN) powder	L. delbrueckii subspp. Bulgaricus, and Streptococcus salivarus subspp. thermophilus	Dietary fiber from Bambara groundnut	• Foam-mat dried BGN powder showed desirable water absorption (1.27 g/g) and water solubility (73.3 100/g) index. • Probiotic viability in reconstituted BGN powder (7.2 log CFU/mL) remained above the minimum recommended limit.	Hardy and Jideani (2019)

This indicates the need for more research on synbiotic legume-based food products owing to the numerous advantages associated with them.

Legumes belong to the *Fabaceae* family, popular for their easy availability, low price, and high nutritional composition in comparison to other staple food groups in different countries worldwide. They are known for their rich protein content along with vitamins, minerals, and many essential amino acids (leucine, lysine, etc.). For these reasons, legume-based food products are a great choice of food for meat and dairy substitutes. Among all legumes, soybean (*Glycine max*) has been the most popular and widely consumed legume globally. The aqueous extract of soybean, popularly known as soymilk, is the primary milk substitute for vegan or lactose-intolerant people (Table 5.1). Recently, the application of synbiotics has been introduced in soymilk to develop synbiotics in soymilk, yogurt, and instant powder. However, a significant proportion of the population also faces soy allergies. Moreover, the presence of antinutritional factors (ANFs) like trypsin inhibitors also impacts the overall nutritional quality of the product developed using soybeans. Thus, for such consumers, other legumes with similar nutrient profiles can be used to develop non-dairy legume-based synbiotic food and beverages. Such legumes include chickpeas (*Cicer arietinum*), kidney beans (*Phaseolus vulgaris*), peas (*Pisum sativum*), lentils (*Lens culinaris*), green gram (*Vigna radiata*), etc.

5.1.1 Soybeans

Soybean is a high-quality protein-dense legume with no cholesterol or lactose. The major prebiotics in soybean, often called 'soybean oligosaccharides,' include raffinose and stachyose. The presence of these sugars allows the growth of probiotics like *Bifidobacterium* spp., which ferment these sugars to produce desirable metabolites. Similarly, *L. acidophilus* has also been shown to break down soybean oligosaccharides via α-galactosidase activity. As shown in Table 5.1, several recent studies have considered soy-based products such as milk, yogurts, and instant powders as a suitable matrix to carrier probiotics, prevent them during GI transit, and serve as a source of food to them. Nonetheless, the selection of an appropriate probiotic strain as per their survivability in the human gut must be carried out to ensure maximum colonization till the end of the shelf life of the food product. Another reason why soy-based product, especially soymilk, is considered by manufacturers is that it is an excellent source of bioactive peptides. Moreover, fermentation by probiotics enhances the generation of these substances, which in turn tends to show both angiotensin-converting enzyme (ACE) inhibitory activity and antioxidant properties. Studies have also shown the capability of synbiotic soymilk in improving oxidative stress factors in patients with diabetic kidney disease. Hence, the synbiotic fermented soybean-based products present a great innovative functional food with desirable sensory acceptability and numerous health benefits.

5.1.2 Chickpeas

Chickpea is among the oldest and most widely consumed legumes in various parts of the world. It is an excellent source of essential amino acids, resistant starch, raffinose-family oligosaccharides (RFOs), and fibers. Chickpea grains are rich in α-galactooligosaccharide (α-GOS), consisting of ciceritol and stachyose, which act as prebiotics in functional foods, owing to their ability to be fermented by colonic bacteria and modulate the intestinal microbial diversity to promote gut health in humans. Chickpea consumption also possesses other health benefits, such as the prevention of hypocholesterolemic and cardiovascular diseases. It also exhibits anti-cancerous, anti-diabetic, and anti-inflammatory activity. Nonetheless, the consumption of chickpeas is often related to the problems like bloating, stomach ache, and flatulence due to the partial breakdown of oligosaccharides or other antinutrients. Hence, these legumes are processed before consumption, either by germinating, roasting, or fermenting.

The functional food market includes various chickpea-based food products, such as chickpea milk as a dietary option. However, the inclusion of fermented chickpea milk to produce synbiotic non-dairy substitutes is still at an infant stage. Scientists have studied and proven the potential of such a product with exceptional properties and health benefits (Table 5.1).

5.1.3 Kidney Beans

Kidney beans, often called common beans, are a part of daily diet worldwide, specifically in Latin America, Eastern Africa, and India. It is consumed throughout the year and is produced in different varieties. Kidney beans are rich in minerals like iron, magnesium, potassium, and zinc. They also have a good amount of carbohydrates in the form of starch and fibers. Numerous health benefits of kidney beans have been attributed to their high phenolic content and antioxidant activities. Several authors have also explored the prebiotic potential of these beans, denoting the sugars from RFOs, namely, raffinose, verbascose, and stachyose, as the major prebiotics. The fermentation of these indigestible sugars with appropriate probiotics allows the production of short-chain fatty acids (SCFAs), organic acids, and other metabolites. Hence, the administration of such synbiotic foods tends to have positive effects on the body, wherein the overall metabolic features improve, and the host's gut health enhances. Various studies have explored the potential of synbiotic kidney bean beverages and by-products like yogurt and instant powder from them, indicating their beneficial effects. Though no such products are available at commercial levels but owing to their benefits, this system serves as an actual field of opportunity in the functional food market.

5.1.4 Other Varieties of Legumes and Blends

Recently, in the functional food industry, there has been a growing inclination toward the use of legumes, either individually or in combination, as prebiotic-rich probiotic carriers to develop synbiotic non-dairy-based food items. The combination of different legumes not only uplifts the nutritional value of the whole product but also enhances the prebiotic content, which in turn allows maximum colonization of probiotic microorganisms, known for several health benefits. Blending different legumes also plays a crucial role in formulating a product with an acceptable sensory profile because some legumes are known for their typical beany flavor, which resists their consumption. For example, synbiotic kidney bean-based beverage is often refused because of its typical beany flavor. However, in combination with other legume-derived beverages like green mung bean, in adequate proportions, they can neutralize and suppress that distinguished taste and improve the overall sensory acceptance of the final product. Similarly, the removal of outer skins or peels can be done to overcome this issue. Several researchers have worked along similar lines to optimize the proportions of different legume-based beverages to obtain a product with desirable sensory properties as well as suitability for fermentation with probiotics to develop a synbiotic product. Similarly, lentil and kidney bean sprouts have shown effective carrier properties, specifically for probiotic yeast *Saccharomyces cerevisiae* var. *boulardii*. Further, the two legume-based matrices protect the probiotics during gastric digestion and allow maximum growth by providing the required prebiotics to them. Similarly, the probiotic *L. plantarum* has been shown to enhance the starch digestibility of the beverages made using a blend of sprouted lentils and mung beans. Hence, combining two or more legumes coupled with sprouting exhibits an efficient food medium for adequate probiotic delivery and their subsequent growth. Such a synbiotic system can be utilized to improve the microbial load, nutrient digestibility, and overall nutritional composition of these products.

5.2 Processing of Pro/Synbiotic Legume-Based Food Products

After the selection of suitable raw legumes with desirable organoleptic and nutritional properties, the next step of extraction is of major concern. It is critical in the development of an acceptable plant-based dairy substitute as different extraction processes might affect the composition of a specific legume or pulse differently, which in turn is responsible for determining the behavior of the extract during subsequent processing steps. The stepwise extraction process for the development of pro/synbiotic legume-based food is depicted in Figure 5.2.

5.2.1 Pre-treatment

Before milling and grinding the legumes, different pre-treatments can be applied to them based on the physicochemical composition of the legumes.

5.2.1.1 Roasting Roasting is a kind of thermal treatment that is applied to legumes to increase their flavor and aromatic compounds in the powders, flours, and aqueous extracts obtained from them. Roasting also tends to decrease the associated off-flavors and improve

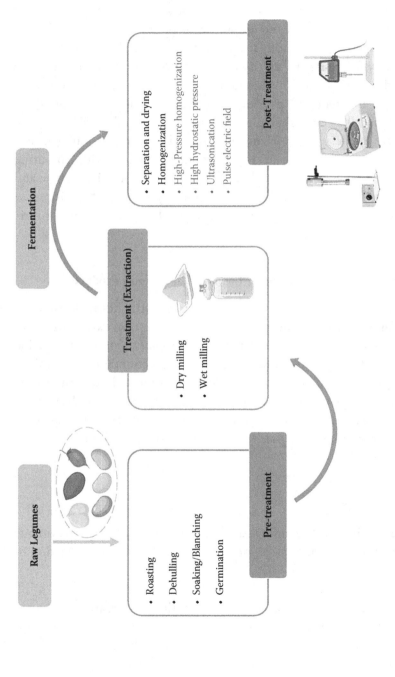

Figure 5.2 Stepwise extraction process for the development of synbiotic legume-based food products.

the overall sensory acceptability by adding 'roasted notes' to the developed product.

5.2.1.2 Dehulling Dehulling is another important step that can be done to a legume or legume-based product based on its desirability. For example, the peels often contribute as a source of prebiotics to potential probiotic microorganisms; hence, dehulling is not encouraged. However, the hulls sometimes impart undesirable compounds to the developed product, causing rejection of the product. Dehulling can be done either by the dry method or the wet method. This step is known to eliminate dietary fibers in legume powders, which in turn is responsible for a better, and homogenized texture of the powder. Dehulling also diminishes any associated off-flavors and ANFs.

5.2.1.3 Soaking and Blanching Legumes are mostly subjected either to soaking in cold water or blanching in hot water to soften the legume seeds and eliminate any undesirable water-soluble substances and associated off-flavor in them. The added heat from blanching sometimes works better in comparison to soaking by allowing a better reduction of unwanted compounds. The heat treatment also inactivates undesirable enzymes, such as lipoxygenase, which causes off-flavors in soy-based aqueous extracts, and trypsin or protease inhibitors, which alter protein assimilation. In this way, it improves the overall flavor and nutrition. Nonetheless, blanching may sometimes also causes inactivation of desirable nutrients such as folate and choline. Hence, based on the type of legume, selection of either soaking or blanching must be done.

5.2.1.4 Germination Sprouting or germination is associated with the previous step of soaking wherein the enzymes present in the legumes like amylases and proteases, get activated, causing a significant change in the composition and related properties of the legume. Germination also enhances the flavor and sensory acceptability of developed products. It also increases the nutritional value of fermented legume-based food products by increasing the antioxidant activity and reducing the ANFs present in the legumes.

5.2.2 Extraction

The process of extraction or milling of pre-treated legumes produces two major extracts: (1) solid powder/flour-based extract and (2) aqueous-based extracts produced by dry and wet milling, respectively. While the powder-based extracts can be reconstituted later to obtain a beverage, wet milling usually generates a finely suspended solution. Nonetheless, products from both steps require further homogenization to reduce any disparity in the stability and texture of the final product.

Additionally, wet milling is often accompanied by heat treatments, such as cooking the slurry. This is generally done to inactivate ANFs and endogenous enzymes, reduce off-flavor production, and increase the extractability of desirable nutrients, though higher treatment temperatures can alter the other major compounds too.

At the end of the extraction step, the obtained powder or slurry is decanted, subjected to gravity, centrifuged, or filtrated to remove any unwanted coarse particles to avoid the chances of phase separation. Moreover, the extracted product may be subjected to drying to allow convenient transportation, enhanced shelf life, or any other related use.

5.2.3 Fermentation

Biological techniques like fermentation with beneficial microorganisms like probiotics are of great interest to the legume-based functional food industry owing to the numerous benefits exerted by them on human health as well as on the quality of the product. Fermentation is a process that is known to enhance the taste and flavor of a product because of the release of different organic compounds. Fermentation also exhibits higher antioxidant and anti-microbial properties, which in turn have beneficial effects on humans. Consequently, a legume-based fermented product offers a higher shelf life and variety to vegans or those who are lactose intolerant. In this way, along with a better nutritional profile, fermented legumes also serve as a potential prebiotic matrix for the development of a synbiotic food product that can fulfill the needs of diet-specific populations in the world.

Along with biological techniques like fermentation, the legume-based functional food industry also employs certain chemical treatments to increase the efficiency of the developed product. These chemical methods include the following:

pH Treatment: Changing the pH of the soaking or extraction medium of a legume-based product is done often as it allows easy and fast extraction of protein extractability and reduces associated off-flavor development. Altering pH during the separation step is also used to prevent any kind of phase separation.

Enzymatic Treatment: The application of different enzymes in the non-dairy legume-based industry is also carried out for the breakdown of larger particles into smaller sizes. This is also linked with better solubility and, thus, an acceptable mouthfeel. Particularly, enzyme reaction during the fermentation step also allows the development of desirable flavor characteristics.

5.2.4 Enhanced Functionality with Innovative Processing

Due to the processing challenges associated with the development of pro/synbiotic legume-based food products, there is a greater need for innovative processing technologies in the functional food industry. Such technologies hold the potential to overcome the major challenges related to the texture, off-flavor, and ANFs in the developed product. Moreover, an enhanced shelf life is also offered by some of these techniques. These innovative processing technologies include the following:

High-pressure homogenization (HPH) and ultra-HPH (UHPH) are used to improve the stability and organoleptic properties of legume-based liquid extracts. These techniques help in the attainment of desirable globule sizes with required viscosity, and stability, without hampering the sensory properties of the product. In addition, the protein configuration and functional properties like emulsification and foaming abilities also improve. Additionally, the smaller the particle size, the higher the possibility to make a dispersion or suspension. This in turn stabilizes the mixture, thus enhancing the overall acceptance.

Pasteurization and sterilization are the most used methods for preventing pathogenic growth in pro/synbiotic-based products. However, the potential of novel innovative non-thermal techniques has been explored recently by the legume-based functional food industry for carrying out effective pathogenic inactivation. These techniques include ultrasonication, high hydrostatic pressure (HHP), and pulsed electric field (PEF) treatment.

5.3 Fermented Pro/Synbiotic Legume-Based Beverages

Since beverages provide a better medium for the delivery of probiotics, the functional food industry has explored the potential of various legumes and pulses in the development of synbiotic legume-based beverages, which can be utilized as a dairy substitute. Numerous researchers have worked on formulating non-dairy pro/synbiotic beverages based on legumes as the delivery matrix for probiotics as well as a source of prebiotics. In this way, legume based beverages serves the purpose of synergism by creating a synbiotic environment. As discussed earlier, the richness of legumes in non-digestible oligosaccharides provides an energy source to the probiotic microorganism, creating a healthy food product.

The most common legumes used by the food industry to develop synbiotic legume beverages include soybeans and probiotics like *Bifidobacterium animalis* Bb-12 and *L. acidophilus* La-5. Probiotic-fermented soybean milk serves as a protein-rich substrate and exhibits proven abilities to decrease the occurrence of osteoporosis and menopausal symptoms. Several studies have shown the potential of soybeans as a suitable probiotic delivery matrix by allowing desirable viability and tolerance in the GI environment. Sometimes, additional prebiotics like inulin as well as process by-product like okara flour are often added to probiotic soymilk to enhance the viability of prebiotics and create a more sustainable synbiotic system. Recent studies are now working to improve the overall flavor and texture of the pro/synbiotic soy drinks as, in comparison to their dairy counterparts, soy-based beverages lack an acceptable taste. For this purpose, probiotic soy drinks are often combined with cereals (germinated wheat, millet, barley flour), legumes (green mung bean, kidney bean), or nuts (peanut) to produce beverages with better sensory acceptance, enhanced probiotic properties, and higher viability throughout the shelf life. Such fortification methods also have additional health benefits, such as modulation of the gut microbiome via increasing beneficial microbes and decreasing pathogens, enhancement of glucose metabolism, and protection against metabolic alterations of diabetic pathology.

Nonetheless, a considerable section of the global population suffers from soy allergies and resists the intake of soybean and soy-related food products. This creates a need for more variety in non-dairy products, particularly for people who have soy allergies or those who follow a vegan diet. For this purpose, researchers have explored the potential of other legumes to produce synbiotic legume beverages with added health benefits with a major focus on improved gut health. Such legume-based beverages include chickpea, green mung bean, and red kidney bean-based synbiotic beverages. Studies have shown the potential of these legumes to serve as a potential delivery matrix along with a rich source of prebiotics (oligosaccharides). Such legumes can be used individually or in combination to produce a synbiotic beverage. Moreover, drying and formulating ready-to-reconstitute instant beverage powder using these legumes have also been carried out. Such products provide convenience to consumers along with longer shelf life and additional health benefits by both the components. Such non-dairy synbiotic legume-based beverages have shown the ability to reduce the cases of type 2 diabetes and high blood pressure. Additionally, these beverages also exhibit good sensory acceptability with the potential to be used by soy-allergic, vegan, or health-conscious people as a dairy substitute.

5.4 Summary

In the modern functional food industry, the addition of probiotic microorganisms in combination with prebiotic substances to develop a synbiotic food has been accepted by consumers.

Among the different food groups, legumes are a rich source of compounds that tend to exhibit prebiotic properties.

The major prebiotic substances in legumes include RFOs, α-GOS, etc., which act as a selective substrate for specific probiotic microorganisms.

The probiotics thrive and grow in the presence of these legume-derived prebiotics and exhibit several pro-health properties.

The synbiotic legume-based food products are widely accepted by the population who suffers from lactose intolerance, milk allergies, or is inclined to veganism and requires more dietary options.

The utilization of legumes not only serves as a healthy food option but is also associated with several combined benefits, particularly the improved gut health and an enhanced immune system.

The use of nutrient dense under-utilized legumes must get necessary attention for developing legume-based synbiotic products.

5.5 Multiple Choice Questions

1. Which among the following is lesser explored food group with respect to synbiotic food?
   a. Cereals
   b. Legumes and pulses
   c. Fruits and vegetables
   d. Dairy
2. Advantages associated with synbiotic legume-based beverages are ____.
   a. added nutrition and health benefits
   b. potential milk substitute
   c. food option for lactose-intolerant, vegan people
   d. All of these
3. Soybean oligosaccharides include ____.
   a. raffinose
   b. stachyose
   c. Both a and b
   d. None of these
4. Chickpea grains are rich in _____ consisting of ciceritol and ___.
   a. α-galactooligosaccharide; stachyose
   b. fructose-oligosaccharides; raffinose
   c. α-galactooligosaccharide; raffinose
   d. fructose-oligosaccharides; stachyose
5. Legume(s) used to make synbiotic beverages for soy-allergic, vegan people is/are ____.
   a. mung bean
   b. kidney bean

    c. chickpea

    d. All of these

6. Flavor of soybean-based synbiotic beverages can be improved by _____.

    a. fortifying with additional prebiotics like inulin

    b. fortifying with process by-products like okara flour

    c. blending with different cereals, legumes, and nuts

    d. All of these

7. Which of the following is NOT a pre-treatment method for the development of synbiotic legume-based food products?

    a. Fermentation

    b. Roasting

    c. Soaking

    d. Dehulling

8. Extraction or milling of pre-treated legumes produces _____.

    a. solid powder/flour-based extract

    b. aqueous-based extracts

    c. Both a and b

    d. None of the above

9. Biological technique essential for the preparation of synbiotic legume-based food products is _____.

    a. pH treatment

    b. enzymatic hydrolysis

    c. fermentation

    d. All of these

10. Which of the following is NOT an innovative technique used in synbiotic legume-based industry?

    a. High-pressure homogenization

    b. High hydrostatic pressure

    c. Pulse light treatment

    d. Ultrasonication

5.6 Short Answer Type Questions

Q1. Enlist the different legumes used in the synbiotic industry. Explain any one in detail.

Q2. Write a short note on recent developments in the field of synbiotic legume-based products.

Q3. Differentiate between soaking and blanching in the pre-treatment steps of legume processing.

Q4. What are the biological and chemical treatment methods used during the processing of pro/synbiotic legume-based food products? Describe in short.

Q5. Write a short note on fermented pro/synbiotic legume-based beverages for soy-allergic, vegan people.

5.7 Descriptive Questions

Q1. Discuss the prebiotic and probiotic combinations used for any three legume-based synbiotic products.

Q2. Explain the recent studies done in the field of synbiotic legume-based products highlighting the product, probiotic and prebiotic components, and the key conclusions.

Q3. Discuss in detail: fermented pro/synbiotic legume-based beverages.'

Q4. Elaborate in detail the stepwise processing of a pro/synbiotic legume-based food product.

5.8 Answers for MCQs

Q1	Q2	Q3	Q4	Q5	Q6	Q7	Q8	Q9	Q10
b	d	c	a	d	d	a	c	c	c

References

Chaturvedi, S., & Chakraborty, S. (2022a). Comparative analysis of spray-drying microencapsulation of *Lacticaseibacillus casei* in synbiotic legume-based beverages. *Food Bioscience*, *50*(PB), 102139. 10.1016/j.fbio.2022.102139

Chaturvedi, S., & Chakraborty, S. (2022b). Optimization of extraction process for legume-based synbiotic beverages, followed by their characterization and impact on antinutrients. *International Journal of Gastronomy and Food Science*, *28*, 100506. 10.1016/J.IJGFS.2022.100506

Chen, Y., Zhang, H., Liu, R., Mats, L., Zhu, H., Pauls, K. P., Deng, Z., & Tsao, R. (2019). Antioxidant and anti-inflammatory polyphenols and peptides of common bean (*Phaseolus vulga* L.) milk and yogurt in Caco-2 and HT-29 cell models. *Journal of Functional Foods*, *53*, 125–135. 10.1016/J.JFF.2018.12.013

Falade, K. O., Ogundele, O. M., Ogunshe, A. O., Fayemi, O. E., & Ocloo, F. C. K. (2015). Physico-chemical, sensory and microbiological characteristics of plain yoghurt from bambara groundnut (*Vigna subterranea*) and soybeans (*Glycine max*). *Journal of Food Science and Technology*, *52*(9), 5858–5865. 10.1007/S13197-014-1657-3/TABLES/3

Hardy, Z., & Jideani, V. A. (2019). Functional characteristics and microbiological viability of foam-mat dried Bambara groundnut (*Vigna subterranea*) yogurt from reconstituted Bambara groundnut milk powder. *Food Science and Nutrition*, 1–11. 10.1002/fsn3.951

Jia, M., Luo, J., Gao, B., Huangfu, Y., Bao, Y., Li, D., & Jiang, S. (2023). Preparation of synbiotic milk powder and its effect on calcium absorption and the bone microstructure in calcium deficient mice. *Food and Function*, *14*(7), 3092–3106. 10.1039/d2fo04092a

Murevanhema, Y. Y., & Jideani, V. A. (2020). Shelf-life characteristics of Bambara groundnut (*Vigna subterranea* (L.) Verdc) probiotic beverage. *African Journal of Science, Technology, Innovation and Development*, *12*(5), 591–599. 10.1080/20421338.2020.1767354

Rajasekharan, S. K., Paz-Aviram, T., Galili, S., Berkovich, Z., Reifen, R., & Shemesh, M. (2021). Biofilm formation onto starch fibres by *Bacillus subtilis* governs its successful adaptation to chickpea milk. *Microbial Biotechnology*, *14*(4). 10.1111/1751-7915.13665

Riasatian, M., Mazloomi, S. M., Ahmadi, A., Derakhshan, Z., & Rajabi, S. (2023). Benefits of fermented synbiotic soymilk containing *Lactobacillus acidophilus*, *Bifidobacterium lactis*, and inulin towards lead toxicity alleviation. *Heliyon*, *9*(6). 10.1016/j.heliyon.2023.e17518

Sengupta, S., Koley, H., Dutta, S., & Bhowal, J. (2019). Hepatoprotective effects of synbiotic soy yogurt on mice fed a high-cholesterol diet. *Nutrition*, *63–64*, 36–44. 10.1016/J.NUT.2019.01.009

6

SYNBIOTICS IN FRUIT AND VEGETABLE INDUSTRY

The fruit and vegetable (F&V) industry is one of the most important industries when discussing non-dairy food matrixes. The main reason for its popularity is the sensory acceptability of the products made using various fruits and vegetables along with the well-known health benefits. Moreover, the addition of probiotics to F&V-based products, especially fruit juices, has increased the demand of health-conscious consumers. F&V exhibits potential as a source of prebiotics, being a rich source of carbohydrates such as fiber, inulin, and oligofructose, fructooligosaccharide (FOS), and galactooligosaccharides (GOS). Additionally, the use of their by-products such as peels and pomace also serves as a rich prebiotic source that can be utilized to develop different innovative synbiotic products. In the last few years, the demand for functional foods has increased drastically, with a particular focus on products comprising prebiotics and probiotics, either individually or in combination (also known as synbiotics). F&V-based synbiotic products have a primary influence on the growth, development, and maintenance of the host's gut health. Hence, the food industry needs a smart and innovative use of F&V and their by-product for the development of synbiotic products with desirable physicochemical properties and longer storage life.

6.1 F&V for the Development of Synbiotic Products

In the non-dairy industry, fruits, cereals, vegetables, and soybeans are the most explored and studied matrices that have shown synbiotic delivery in humans. Among these, fruits and vegetables are considered nutrient-rich healthy foods with the ability to serve as an ideal medium for functional ingredients to exhibit numerous health benefits. F&V consists of various phytochemicals, minerals, vitamins,

 DOI: 10.1201/9781003304104-6

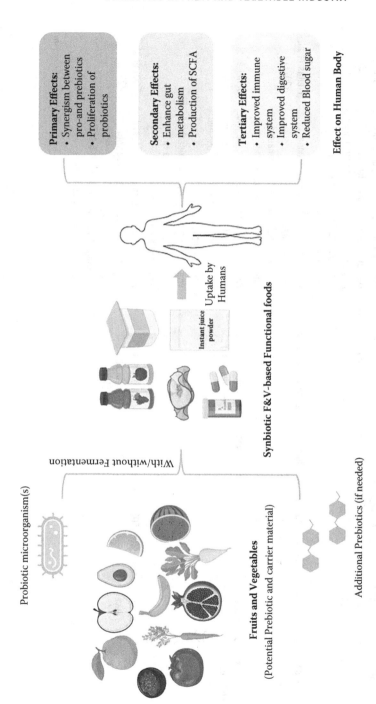

Figure 6.1 Health benefits of synbiotic F&V-based food products in humans.

Table 6.1 Recent Studies Carried Out for the Development of Synbiotic F&V-Based Food Products

FRUIT/VEGETABLE	PRODUCT	PREBIOTIC	PROBIOTIC	KEY CONCLUSION	REFERENCE
Pomegranate	Synbiotic pomegranate juice	Xylooligosaccharides (XOS) from sugarcane bagasse	*Bifidobacterium animalis* Bb-12	• Product had a probiotic viability of 10^6 CFU/mL even after 30 days of cold storage. • XOS allowed protection of probiotic bacteria in the low acidic environment of pomegranate.	Hesam et al. (2023)
Açaí (*Euterpe oleracea*)	Synbiotic açaí juice	Fructooligosaccharides (FOS) and sucrose	*Lacticaseibacillus casei*	• Fermentation enhanced the functional value of the beverage. • *L. casei* viability reduced but remained within the minimum required limit.	Freitas et al. (2021)
Yacon (*Smallanthus sonchifolius*)	Synbiotic yacon juice	Inulin and FOS in yacon	*Lactobacillus acidophilus* (La5)	• Probiotics maintained viability in the juice at a satisfactory level (10^6 CFU/mL) for 27 and 15 days, respectively, in refrigerated and room conditions. • Vitamin C, phenolics, and antioxidant activity were comparatively more in fermented juice to control juice.	Dahal et al. (2020)
Apple	Synbiotic apple juice	FOS in apple	*Lactobacillus plantarum* ATCC 8014	• 2.3% FOS, 213 mg/L ascorbic acid, and 1.4 g/L citric acid in apple juice with 3.6×10^{11} CFU/mL probiotic improved the efficiency of patulin removal to 95.91% during six weeks of cold storage.	Zoghi et al. (2019)

(Continued)

Table 6.1 (Continued) Recent Studies Carried Out for the Development of Synbiotic F&V-Based Food Products

FRUIT/VEGETABLE	PRODUCT	PREBIOTIC	PROBIOTIC	KEY CONCLUSION	REFERENCE
Grape	Synbiotic grape Juice	Oligofructose and polydextrose	*Lacticaseibacillus casei*	• There was no significant difference in flavor and color of optimized synbiotic apple juice until three weeks of cold storage. • Pro/prebiotics did not alter the physicochemical properties or storage stability of the juices. • Probiotic viability remained higher than 10^6 CFU/mL with sensory acceptance the same as that of the control juice.	Silva et al. (2021)
Cornelian cherry (*Cornus mas* L.)	Synbiotic fermented Cornelian cherry beverage	Delignified wheat bran	*Lactobacillus paracasei* K5	• Total phenolics content (TPC) during the whole cold storage period, was higher in comparison to control juice. • Probiotic viability remained at higher levels (9.74 log CFU/mL) till the fourth week of storage.	Mantzourani et al. (2019)
Orange and carrot	Synbiotic orange-carrot juice	Burdock root (*Arctium lappa* L.) encapsulated extracts	*Lactobacillus acidophilus*	• Juices fortified with encapsulated extract significantly improved probiotic viability, TPC, and formalin index during cold storage. • Addition of root extract did not have an adverse effect on the flavor and odor of the samples and showed acceptable sensory scores (6.9–8.8).	Esmaeili et al. (2022)

antioxidants, enzymes, and dietary fibers. They also contain high sugar percentages, which makes them an important probiotic carrier for allowing maximum growth of beneficial microorganisms.

In comparison to dairy products, F&V does not contain any allergic substances like lactose or cholesterol, which have shown adverse effects in certain population groups, such as lactose-intolerant people or those with milk allergies. Hence, F&V-based synbiotic products can be utilized to serve such a population of society to prevent them from depriving essential nutrients. Moreover, F&Vs are fresh and healthy, with an appealing mouthfeel that is known to please a majority of all age groups in a population. All these features allow researchers and commercial food industry experts to explore the area of synbiotic F&V-based products. Though the literature is rich in studies with probiotic-based F&V-based juice, the potential synergistic effect of the two components (prebiotics and probiotics) must be explored in depth to understand the favorability and wide acceptability of such a product by the consumers (Figure 6.1). A variety of probiotic F&V products have been developed and commercialized in the food industry, which includes juices, dried fruits, and fermented vegetables, ready-to-serve instant powder. Similarly, studies are also going on synbiotic F&V-based products, as shown in Table 6.1.

A majority of common as well as underexplored F&Vs as per their regional cultivation in different countries, have been studied for the development of pro/synbiotic food products. Apples, oranges, mangoes, beetroot, garlic, onion, artichoke, cabbage, and tomatoes are the most used F&V for making synbiotic products like juice, yogurts, and fermented products. Similarly, a wide range of probiotic strains, mainly species of *Lactobacillus* and *Bifidobacteria*, such as *Lactobacillus acidophilus, L. plantarum, L. rhamnosus GG, L. casei, L. paracasei, L. fermentum, and B. bifidum,* are used in the processing of probiotic or synbiotic F&V-based food items.

6.2 F&V-Based Products

The presence of lactose and cholesterol content in probiotic dairy products creates a setback for its commercialization. In this sense, technological advancements in the form of non-dairy substitutes using F&V as a potential delivery matrix allow a scope to broaden the horizons of the functional food industry with improved probiotic viability and functionality. F&V can be modified as per the desired requirement by either adding the required food additives or removing undesirable antinutrients. In this way, F&V serves as an ideal substrate for probiotic microbes having a nutrient-rich composition, without any dairy-based allergens. Additionally, the pleasing and refreshing taste profile offered by these food groups is a boon with maximum consumer interest and acceptance by all age groups.

Probiotic and synbiotic-based fruit products are the most suitable sources for the growth and survival of probiotics. This is because of their rich nutritional profile with high contents of vitamins, minerals, antioxidants, phytochemicals, and dietary fibers, which serves as a source of energy and support the growth of probiotics. Among these, probiotic and synbiotic fruit juices are the most popular and widely accepted products among all age groups being healthy, energizing, and available in plenty of flavors. The most common fruit-based pro/ synbiotic uses apples, pineapple, strawberry, sweet lime, mangoes, olives, oranges, etc. Several traditional probiotic fruit-based juices are available in the market. For example, 'Hardaliye' is a traditional beverage in Turkey made by fermenting grapes via lactic acid bacteria, probiotic *Lacticaseibacillus casei* subsp. *pseudoplantarum* and *L. paracasei* subsp. *paracasei,* along with crushed mustard seeds and a few proportions of benzoic acids. Moreover, the pro/synbiotic fruit juices are also often blended with herbs such as green tea (*Camellia sinensis*), Ashwagandha (*Withania sominifera*), wheatgrass, and aloe vera to make a healthier version of synbiotic phytochemical-enriched mix fruit juice variant.

Likewise, vegetable-based pro/synbiotic beverages are also rich in essential nutrients that help in the maintenance of the overall functioning and homeostasis of the body (James, Wang, 2019). Vegetables like carrot, tomato, beet, onion, bamboo shoots, potato, ginger, broccoli, eggplant, mustard leaves, cabbage, and cauliflower are the most explored ones for the development of non-dairy vegetable-based synbiotic products. Some of the most commonly available traditional fermented foods made using vegetables include kimchi, kombucha, sauerkraut, etc., which are consumed in Korea, Japan, and China. Similarly, Kanji is an Indian probiotic beverage made using carrots and beets and the probiotic *Lactobacillus acidophilus* for the fermentation process. Several F&V juices are also blended along with the addition of external carbohydrates to increase the overall prebiotic activity.

Thus, F&V serves as a potential non-dairy probiotic delivery matrix with high prebiotic activity and can be efficiently utilized for the development of synbiotic food products. Moreover, F&V juices serve as the most convenient food option for the delivery of probiotics with high nutritional value and sensory acceptability and longer shelf life with added health benefits.

6.3 Probiotics as Starter Culture in F&V-Based Synbiotic Products

The ability of probiotic microorganisms to grow and flourish in fermented F&V has been studied and proven in the literature extensively (Rasika et al. 2021). Studies have shown the successful use of commercial probiotic culture in F&V-based products in the development of synbiotic food products. A food system needs probiotic starter culture or an initial inoculation microbe for standardized and uniform formulations of products at the commercial level. In the food industry, the major probiotics used for the development of synbiotic food products include *Lactobacillus acidophilus, L. casei, L. paracasei, L. bulgaricus, L. lactis, L. plantarum, L. reuteri, L. fermentum, L. brevis, L. rhamnosus, B. longum, B. bifidum, B. infantis, B. breve, B. animalis, B. lactis, Streptococcus thermophilus, Saccharomyces cerevisiae,* etc. These probiotics can be used individually or in combination based on the type of carrier material and prebiotic to exert maximum probiotic and prebiotic potential.

Moreover, to ensure the highest possible probiotic activity without any disconformities during the processing steps, the selected probiotic for a specific medium must follow some prerequisites as described below:

1. General probiotic properties: The source, pathogenicity, and toxicity of the strain should be known. The ability of antibiotic resistance, resistance toward gastric pH and intestinal bile salts, presence of antimicrobial components, and required acidification rate must be known. The strain should have GRAS (generally recognized as safe) status.

2. Properties during processing: The strain must endure different processing conditions like fermentation, variation in temperatures, oxygen levels, pH, and moisture content. It should also remain viable during storage.
3. Sensory profile: Ability to produce desirable aroma composites, and other flavor compounds from lactic acid accumulation or via proteolytic and lipolytic enzymatic reactions.
4. Functional properties: Ability to produce biogenic complexes, must allow an increase in antioxidant activity, hydrolyze exopolysaccharides, etc.

In some F&V, even a minor change in environment can alter the growth of a specific strain, which in turn can cause varying fermentation products, metabolite substrates, and their composition, and eventually, the acidity, texture, flavor, and overall composition of the developed synbiotic product will be hampered. Thus, it is important to sustain a preserved atmosphere for maintaining the functional and sensory quality of pro/synbiotic F&V-based products.

6.3.1 Preservation and Antagonistic Properties of Probiotics

Conventionally, lactic acid bacteria or LAB have been used in most of the F&V processing and preservation methods to exhibit fermentation and produce related advantageous compounds to improve the overall quality of the developed product. Different LAB species produce different antagonistic biochemicals which function against the growth of pathogenic and spoilage-causing microbes in F&V-based products. These antagonistic biochemicals generally include primary and secondary metabolites such as organic acids, hydrogen peroxide, carbon dioxide, ethanol, acetaldehyde, antifungal compounds, free fatty acids, bacteriocins, antibiotics, etc.

However, the recent use and incorporation of probiotics in F&V-based food products has not just enhanced the overall preservation process of the product but has also improved the consumer's demand for fresh, healthy, and long-lasting food alternatives. Among the various metabolites produced during probiotic fermentation of F&V-based products, bacteriocins are of great interest in allowing successful inhibition or reduction of pathogens. For example, studies have shown the ability of bacteriocins like enterocin to reduce the viability of *Listeria monocytogenes, Pediococcus parvulus, Bacillus cereus, Staphylococcus aureus, Lactococcus collinoides,* and *Lactobacillus diolivorans*. Similarly, the combination of metabolites like enterocin and nisin allows the inactivation of the spoilage bacteria like *Alicylobacillus acidoterrestris*. Thus, the application of specific probiotic strains with such preservative and antagonistic properties is crucial and can be considered while formulating an F&V-based food product.

6.4 Synbiotic F&V-Based Food Product: Challenges and Their Possible Solutions

6.4.1 Factors Affecting Survivability of Probiotics in a Synbiotic F&V-Based Food Product

Though F&V are nutrient-rich food commodities, processing them into different products such as juices, syrups, and dehydrated slices often reduces the essential peptides and free amino acids, which in turn causes limited energy sources for probiotic culture for their growth. Similarly, several other factors influence the stability of probiotics in F&V matrices. These include the type of microorganism strain and its inoculum concentration; pH, acidity, and water activity; antimicrobial compounds, additives, and antinutritional factors of the fruit or vegetable or the product thereof; the production processes (pasteurization/cooking time-temperature); and consequent product storage and handling conditions (storage temperature, packaging material, and sterile environment).

pH of the fruit/vegetable/product: Among these, one of the main factors that plays a crucial role in the survivability of probiotics in a synbiotic F&V-based food product is its pH. For a long time, F&V-based product manufacturers, mainly dealing with pro/synbiotic juices and beverages, have faced challenges in controlling the physicochemical parameters of juices where the F&V with lower pH fails to attain the required total viable counts of probiotics. Though fruits are considered the desirable matrix for probiotic growth, the survivability of these microbes is more complex in comparison to dairy products. This is because of the acidity of fruit juices, caused by higher concentrations of organic acids, which decrease the pH and affect probiotic bacteria by disrupting their cell wall. Probiotic cells are not able to thrive in a low pH matrix as the energy provided by ATP for the preservation of the intracellular pH also increases, which in turn causes depletion of ATP cells, which has a negative effect on the related functioning of cells. Moreover, the lack of peptides and free amino acids in F&V juices often disrupts the metabolism of probiotic microorganisms and eventually reduces their viability during the shelf life of the product.

Probiotic strain: In addition to the low pH, the probiotic viability in an F&V product also depends on the specific strain of probiotic in a particular matrix, i.e., the type of fruit or vegetable used. This indicates the exclusivity of probiotics with respect to the matrix, which in turn is responsible for determining the acceptability as well as endurance of the product during the storage period. As mentioned earlier, probiotics belonging to the *Lactobacillus* spp. can endure and grow exponentially in a pH range of 4.3 to 3.7. On the other hand, *Bifidobacteria* strains are less tolerant even at a pH of 4.6. Nonetheless, this trend is specific to probiotics, or it depends on the presence of specific components in a fruit or vegetable matrix. For example, in some fruit juices, ingredients like ascorbic acid, which is known to decrease the redox potential; saccharides or organic acids, which are used as a carbon source; or cellulose, which guards probiotics from harsh processing and storage conditions, help to sustain the required viability of probiotic microorganisms irrespective of their acidic pH.

In addition to this, a developed synbiotic F&V-based product must also deliver the required health benefits via an adequate nutritional profile and sensory acceptance by consumers for convenient commercialization. Hence, these factors create major challenges in the formulation of synbiotic F&V-based food products. Though the functional food industry is working on the best possible solutions to overcome these challenges, the consumers too need to be convinced for the consumption of such products.

6.4.2 Possible Solutions to Overcome the Challenges Faced by Synbiotic F&V-Based Industry

Synbiotic F&V-based food products like juices or beverages are not among the commonly available commodities on the commercial front.

Rather, this sector represents itself as a small niche part of the market. The reason for being skeptical about the acceptance of such products by consumers is the presence of live microorganisms in them, which has also not been strictly regulated by food authorities and regulatory bodies. Apart from this, from a technical point of view, the survival of probiotics in the developed F&V-based synbiotic product during the processing and storage is another crucial issue that determines its acceptance. As discussed earlier, as per FAO/WHO guidelines, a probiotic product must have a minimum of 10^6 colony-forming units (CFU) per gram or milliliters in the food product. In this way, any synbiotic food product must also confer this minimum viable count still the storage period.

Thus, in order to maintain and eventually enhance the total viable count of probiotics in a synbiotic F&V-based food product, different solutions have been implemented. These include (1) fortification with prebiotics and other food additives and (2) microencapsulation of probiotic microorganisms.

6.4.2.1 Fortification with Prebiotics and Other Food Additives The addition of prebiotics along with other food additives could be considered a solution to processing challenges faced by synbiotic products. Several studies have worked on improving probiotic survivability in different F&V-based matrices. Among those, fortification by prebiotics such as dietary fiber and cellulose has proven its capacity to confer good protective mechanisms in synbiotic food products. This is because of the biased adaptation of probiotics toward the prebiotic components that allow their enhanced growth. Now, a prebiotic, along with other food additives, can be added to a synbiotic F&V-based juice either directly to the juice or by allowing fermentation of the juice with added prebiotics, food additives, and the probiotic microbe. In general, carrying out fermentation is considered more beneficial than the direct addition as numerous microbial postbiotic metabolites such as bacteriocins, peptides, and phenols are produced, which enhances the quality and shelf life of the product.

6.4.2.2 Microencapsulation of Probiotic Microorganisms Maintenance of desirable pH in F&V juices is one of the main challenges faced by the synbiotic industry. This is because of the tolerance ability of specific probiotic bacteria in F&V-based matrices. Hence, for this purpose, microencapsulation of probiotics or the mixture of prebiotics and probiotics is carried out using various polysaccharides and protein-based coating materials (Rovinaru, Pasarin, 2019). In this way, the probiotic viability can be protected as well as preserved for a longer duration. Various studies have also shown the ability of encapsulated synbiotic mixtures to endure the harsh conditions of the GI tract in humans. However, irrespective of the great potential exhibited by F& V-based food products as a carrier material for microencapsulated pro/synbiotics, there is very limited literature available. There are limited studies that have explored the physicochemical and micro-biological properties of developed encapsulated synbiotic substrates along with the shelf-life analysis and consumer suitability. Nonetheless, future work could also be carried out to explore the prebiotic potential of polysaccharides, which are used as encapsulating agents.

6.5 Synbiotic System from F&V-Based Waste

In recent times, the F&V industry, which is known to generate the highest amount of agro-industrial waste, has employed the aspects of valorization. The occurrence of bioactive substances and major phenolic compounds in the peels, pomace, and other related waste from F&V allows a scope of smart and efficient usage for the food industry. Additionally, F&V-based waste is important for valorization in the food industry because of its low price, higher production, and valuable bioactive potential. Several researchers have carried out the extraction of desirable compounds like polyphenols (prebiotics) from the F&V waste and incorporation of the extracted compounds in different food systems. Moreover, the concept of designing pro/synbiotic functional juices and beverages using valorized wastes and microencapsulated probiotics also holds scope for innovation in the emerging functional food market.In addition, the use of smart bottles for enhanced product stability and controlled release of the micro-capsules are also some creative approaches for waste utilization in the F&V industry.

6.6 Summary

The F&V industry exhibits itself as one of the most important sectors owing to its refreshing and flavorful profile with numerous health benefits.

F&V-based food products, specially juices and beverages, are the most consumed food commodities, and they serve as a potential source of prebiotics and allow easy growth of probiotics.

A synbiotic system can be formulated using specific fruit or vegetable, individually or in combination, along with the most suitable probiotic strain(s).

Without any side effects like those shown by dairy products, the lactose- and cholesterol-free composition of F&V-based synbiotic products can be used by lactose-intolerant consumers.

The waste and by-products from the F&V industry like peels and seeds can also be utilized by incorporating probiotics into them and making innovative food options.

Another possibility for the application of synbiotics in the F&V sector involves the use of nanotechnology and genome sequencing that hold the potential to develop smart food with added health benefits.

6.7 Multiple Choice Questions

1. Reason for the popularity of F&V-based synbiotics products is/are _____.
    a. health benefits
    b. refreshing, flavorful, and colorful
    c. acceptable by lactose intolerant and vegan people
    d. All of these
2. Hardaliye is a traditional beverage in Turkey made by fermenting ____.
    a. apple
    b. grapes
    c. pomegranate
    d. orange

3. Synbiotic fruit juices are also often blended with herbs such as ___.
   a. *Camellia sinensis*
   b. *Withania sominifera*
   c. wheatgrass
   d. All of the above

4. Which one of the following is NOT a synbiotic vegetable-based traditional food product?
   a. Kimchi
   b. Kombucha
   c. Wine
   d. Sauerkraut

5. Indian probiotic beverage made using carrots, beets, and the probiotic *Lactobacillus acidophilus* is named as _____.
   a. Soju
   b. Kanji
   c. Kombucha
   d. None of the above

6. A food system needs probiotic starter culture for _____.
   a. cost reduction purposes
   b. making a variety of products
   c. standardized and uniform product formulations at the commercial level
   d. preventing health diseases

7. The probiotic properties that need to be considered during processing are _____.
   a. endures processing conditions like fermentation
   b. tolerates variation in temperatures, oxygen levels, pH, and moisture content
   c. remains actively viable during storage
   d. All of the above

8. Control of acidity and pH of fruit juices for making synbiotic products is essential because ___.
   a. disruption of cell wall of probiotic
   b. depletion of ATP cells
   c. lack of peptides and free amino acids alters the metabolism of probiotics
   d. All of the above

9. Consider the following statements and choose the correct option.
   A. Probiotics belonging to the *Lactobacillus* spp. can endure and grow at a pH range of 4.3–3.7
   B. Probiotics belonging to the *Bifidobacteria* spp. are most tolerant at a pH of 4.6
   a. Statement A is true, B is false
   b. Statement B is true, A is false
   c. Both A and B are true
   d. Both A and B are false
10. Technique to improve the development of synbiotic F&V-based food products is called____.
   a. fortification with prebiotics; microencapsulation of probiotic microorganisms.
   b. freeze drying; addition of MSG
   c. Both a and b
   d. None of the above

6.8 Short Answer Type Questions

Q1. How do fruits and vegetables serve as a potential component of the synbiotic system?
Q2. Enlist the factors affecting the survivability of probiotics in a synbiotic F&V-based food product and explain any one in detail.
Q3. What are the possible solutions to overcome the challenges faced by the synbiotic F&V-based industry? Discuss any one in detail.
Q4. How can waste and by-products from the F&V industry utilized to develop a synbiotic system?
Q5. Discuss in short about the future technologies that could be used for the development of the synbiotic F&V-based industry.

6.9 Descriptive Questions

Q1. Discuss in brief about the role of probiotics as starter culture in F&V-based synbiotic products.

Q2. 'Probiotics in an F&V-based synbiotic system exhibit preservative and antagonistic effects.' Discuss your point of view on this statement.

Q3. What, according to you, is/are the major challenge(s) faced by the synbiotic F&V-based industry? Can you suggest some solutions to it/them?

Q4. Describe the prerequisites for probiotics in a synbiotic F&V-based industry.

Q5. Describe the desirable properties for prebiotics in a synbiotic F&V-based industry.

6.10 Answers for MCQs

Q1	Q2	Q3	Q4	Q5	Q6	Q7	Q8	Q9	Q10
d	b	d	c	b	c	d	d	a	a

References

Dahal, S., Ojha, P., & Karki, T. B. (2020). Functional quality evaluation and shelf life study of synbiotic yacon juice. *Food Science & Nutrition, 8*(3), 1546–1553. 10.1002/fsn3.1440

Esmaeili, F., Hashemiravan, M., Eshaghi, M. R., & Gandomi, H. (2022). Encapsulation of *Arctium lappa* L. root extracts by spray-drying and freeze-drying using maltodextrin and Gum Arabic as coating agents and its application in synbiotic orange-carrot juice. *Journal of Food Measurement and Characterization, 16*(4), 2908–2921. 10.1007/s11694-022-01385-3

Freitas, H. V., Dos Santos Filho, A. L., Rodrigues, S., Gonçalves Abreu, V. K., Narain, N., Oliveira Lemos, T. D., Gomes, W. F., & Fernandes Pereira, A. L. (2021). Synbiotic açaí juice (*Euterpe oleracea*) containing sucralose as noncaloric sweetener: Processing optimization, bioactive compounds, and acceptance during storage. *Journal of Food Science, 86*(3), 730–739. 10.1111/1750-3841.15617

Hesam, F., Tarzi, B.G., Honarvar, M., et al. (2023). Valorization of sugarcane bagasse to high value-added xylooligosaccharides and evaluation of their prebiotic function in a synbiotic pomegranate juice. *Biomass Conversion and Biorefinery, 13*, 787–799. 10.1007/s13399-020-01095-0

James, A., & Wang, Y. (2019). Characterization, health benefits and applications of fruits and vegetable probiotics. *Cyta-Journal of Food, 17*(1), 770–780. 10.1080/19476337.2019.1652693

Mantzourani, I., Nychas, G. E., Alexopoulos, A., Bezirtzoglou, E., Bekatorou, A., & Plessas, S. (2019). Production of a potentially synbiotic fermented Cornelian cherry (*Cornus mas* L.) beverage using *Lactobacillus paracasei* K5 immobilized on wheat bran. *Biocatalysis and Agricultural Biotechnology, 17*, 347–351. 10.1016/j.bcab.2018.12.021

Rasika, D. M. D., Vidanarachchi, J. K., Luiz, S. F., Azeredo, D. R. P., Cruz, A. G., & Ranadheera, C. S. (2021). Probiotic delivery through non-dairy plant-based food matrices. *Agriculture, 11*(7), 599. 10.3390/agriculture11070599

Rovinaru, C., & Pasarin, D. (2019). Application of microencapsulated synbiotics in fruit-based beverages. *Probiotics and Antimicrobial Proteins, 12*(2), 764–773. 10.1007/s12602-019-09579-w

Silva, J. V. de C., da Silva, A. D., Klososki, S. J., Barão, C. E., & Pimentel, T. C. (2021). Potentially synbiotic grape juice: What is the impact of the addition of lacticaseibacillus casei and prebiotic components? *Biointerface Research in Applied Chemistry, 11*(3), 10703–10715. 10.332 63/BRIAC113.1070310715

Zoghi, A., Khosravi-Darani, K., Sohrabvandi, S., & Attar, H. (2019). Patulin removal from synbiotic apple juice using Lactobacillus plantarum ATCC 8014. *Journal of Applied Microbiology*, 126(4), 1149–1160. 10.1111/jam.14172

7

INSTANT SYNBIOTIC FOODS

7.1 Instant Synbiotic Food Products

In recent times, an increase in the trend of healthy lifestyle adaptation by consumers has been observed. In this regard, the use of newer, smarter, and innovative food products is growing enormously owing to the convenience as well as health benefits. Among such products, the wide acceptance of foods and beverages in the form of instant mixes and ready-to-reconstitute food powders has shown their exceptional potential to satisfy consumer's needs. The advantages associated with these products include the longer shelf life, ease in preparation, bulk handling, and transportation. At the same time, incorporation of the concept of synbiotics into the instant powder made the final product healthier, novel, and innovative. Moreover, the dried food powder with added prebiotics and probiotics also helps in the simple absorption of such components in the food products (Chaturvedi et al., 2021). While drying the synbiotic foods into powders reduces the chance of microbial spoilage, it can also alter the activity of probiotics in the product. Hence, the use of optimized drying conditions, drying techniques, and other necessary requirements became a crucial factor in the manufacture of synbiotic instant food powders.

Nonetheless, drying process is an economical and convenient method for preservation of food products employed at the commercial scale to greater customer reach. Thus, drying adds several advantages to the synbiotic instant food powders by conferring (1) extended shelf life, (2) easy bulk handling, (3) flexibility in matrix selection, (4) development of a variety of products, and (5) value addition (Figure 7.1).

Thus, a smart use and selection of suitable matrix from the various groups is essential for developing a synbiotic instant food powder with desirable properties. Different food matrices like dairy, fruits, vegetables, cereals, and legumes in the form of food and beverages

DOI: 10.1201/9781003304104-7

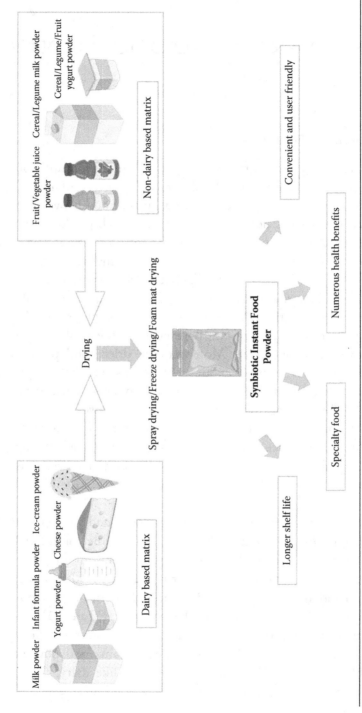

Figure 7.1   Advantages associated with different dairy and non-dairy food matrices for the development of synbiotic instant food powders.

containing probiotics and prebiotics are employed in the food industry for the development of instant functional foods.

7.1.1 Dairy Based

7.1.1.1 Fermented Milk The production and consumption of different synbiotic dairy products such as fermented milk and dairy desserts has been escalating owing to the health benefits they exert. Probiotic-enriched dairy powders such as whole milk powder, skim milk powder, and buttermilk powder are now commonly available. Moreover, recent studies have also involved the concept of using vegetable or tuber components, like beetroot or ginger aqueous extract, as prebiotics in the form of encapsulation to the milk powder. Such products not only increase the efficiency of probiotics during processing and storage but also improve the overall acceptability of the synbiotic fermented milk and milk powders.

7.1.1.2 Yogurt Probiotics have been used in the dairy industry along with prebiotics to make healthy synbiotic yogurts via techniques like co-encapsulation (Rashidinejad et al., 2022). The key aim of this concept is to enhance the survival rate of probiotics in the developed yogurt. Researchers have been trying to advance the healthy functions of yogurts and increase their shelf life by drying and making dried yogurt powders. With such innovation, applications of techniques like fortification make the synbiotic yogurt powder more functional.

7.1.1.3 Cheese Cheese is considered one of the most significant dairy products, with over 1,000 different varieties. In the food industry, apart from its direct consumption, cheese is used as a food ingredient modifying the flavor, mouthfeel, appearance, and other properties. For this purpose, cheese is dried using the spray drying or foam mat drying techniques to produce cheese powder. Such products are of specific interest in the industry to be used as a core ingredient for flavoring or as a nutritional supplement in different food products. The use of probiotic cultures for the manufacture of cheese has also been studied widely and has shown exceptional benefits for human health. At the same time, the incorporation of prebiotic components also allows the scope for the development of vegan cheese. Moreover, the combination of cheese made using probiotic cultures and prebiotic components allows for the development of a synbiotic system, which altogether is a unique concept. Additionally, recent market analyses also show the rapid growth of powdered cheese as an essential ingredient in the food industry.

7.1.1.4 Ice Cream The use of co-encapsulation for producing probiotic and prebiotic-enriched ice cream and frozen desserts has established a distinct interest in recent times. The reason is that co-encapsulation protects probiotic microbes from external stresses caused by the cold treatments of ice milk, which in turn leads to a reduction in the sensory likability of the final product. Moreover, as discussed in previous chapters, the culture medium is an important parameter in maintaining the viability of probiotics in ice cream. The freezing step in ice cream processing leads to the formation of ice crystals inside the cell, which damages the cell wall and membranes. However, this may be prevented by using rapid freezing techniques, which produce smaller ice crystals.

Nonetheless, the concept of making a synbiotic ice cream mix is still at an infant stage in the food industry. Researchers are exploring the possibility of developing a system that consists of a mixture of probiotics and prebiotics together with dairy ingredients, which when reconstituted resembles that of ice cream.

7.1.2 F&V Based

Followed by the dairy sector, F&V-based food items are among the common probiotic carriers. The current trend of consumption of ready-to-reconstitute probiotic beverages in the food industry has also started utilizing fruits and vegetables as their suitable base components. In this regard, extensive research in the literature can be found on different fruits and vegetables as shown in Table 7.1, to optimize the manufacturing of probiotic food powders with acceptable probiotic count along with desirable physio-chemical characteristics, for their effective use at the consumer level. The added advantage exhibited by the synbiotic structure includes the increased ability of certain probiotics like *Bacillus* to tolerate stressful conditions in the host's

Table 7.1 Recent Research on the Development of F&V-Based Synbiotic Instant Powder

PRODUCT	PROBIOTIC USED	PREBIOTIC USED	KEY INFERENCES	REFERENCES
Synbiotic spray-dried litchi juice powder	*Lactobacillus plantarum*	Maltodextrin, fructooligosaccharide (FOS), and pectin	• Powder of maltodextrin 10% (w/v) and 5% (w/v) FOS gave the highest yield and survivability of probiotics. • Powder had good solubility, low water activity, satisfactory color, and uniform particle size with a smooth surface. • Provided protection to probiotics under gastric environment and enhanced their growth in the simulated digestive system.	Kalita et al. (2018)
Synbiotic guava juice powder	*L. plantarum*	Maltodextrin (MD) and inulin (INU)	• Powders stored at refrigerated conditions (4°C) showed higher viability and a longer survival rate up to 45 days. • Better powder yield was obtained by use of 20% maltodextrin. • Maltodextrin with inulin (10% each) showed better survival of synbiotics both while spray drying and storage under refrigerated conditions.	Upadhyay and Dass (2021)
Synbiotic spray-dried orange juice powder	*L. casei shirota*, *L. casei immunitas*, and *L. acidophilus johnsonii*.	Maltodextrins and pectins	• Survival during refrigerated temperature (4°C) resulted in significantly higher counts for all three probiotics up to 60 days. • Combining maltodextrins and pectins at a 10:1 weight ratio showed the best results. • An insignificant reduction in cell viability after the spray-drying process was observed among the three microorganisms.	Gervasi et al. (2022)

GI tract. Such a product can also be employed at the industrial level by serving as a core ingredient for various products. For example, numerous extruded products or masala mixes of different curries use dried vegetables and fruit powders. Additionally, the sector of fruits and vegetables is also used by the non-dairy industry to develop food options for people with specific requirements.

7.1.3 Cereal and Legume Based

The concept of developing synbiotic legume- and cereal-based drinks in the food industry as healthy functional food has proven their potential not just as a non-dairy alternative but also as a nutrient-rich product with various health benefits. Researchers have conducted work on the exploration of various cereals and legumes as sources of prebiotics and combining them with a suitable probiotic micro-organism. However, the strategy of converting the beverage into ready-to-reconstitute drinks or ingredients is still new. The majorly explored cereal- and legume-based products include synbiotic finger millet-based yogurt-like beverages, synbiotic fermented beverages made using aqueous extract of quinoa (*Chenopodium quinoa* Willd), synbiotic soymilk, synbiotic red kidney bean and green mung bean beverage, Bambara groundnut (*Vigna subterranea*) yogurt, and so on. The future possibilities that can be employed in these studies include the scope of drying via spray drying, freeze drying, or foam-mat drying to allow the development of dried powder products that hold the potential to be used as a ready-to-reconstitute instant functional food with added health benefits incorporated via synergistic effects between the probiotic and prebiotic components.

Thus, the smart use of cereals and pulses as a probiotic delivery matrix for manufacturing powdered instant food items is within the space of innovation. Such innovations can provide improved choices for population that faces problems related to lactose intolerance, milk allergies, and cholesterol issues or who, in general, are vegan or vegetarian.

7.1.4 Food Waste Based

Mostly considered waste, the by-products from the agriculture and food industries have gained importance lately. These by-products from processed food waste are rich in numerous secondary metabolites such as anthocyanins, phenolic acids, and flavonols. Even the extraction of some compounds present in the peels and skins of some agricultural waste can be utilized for their functional properties like good water holding, gel formation, prebiotics, and antioxidant abilities. Various studies have utilized by-products like phenolic-rich pomegranate peel, beetroot peel, and grape pomace in the development of a synbiotic system by encapsulating gums and probiotics followed by drying. Such products not just serve as a non-dairy food option with added health benefits but also represent a potential sustainable environment-friendly functional food product. Such products also have a high shelf life, are loaded with nutrients, and are easy to use. Hence, the use of food and agricultural waste in the field of pro/synbiotic foods and beverages in the form of a food matrix acting as a prebiotic source is noteworthy. Nonetheless, more food waste options can be explored in the non-dairy synbiotic beverage industry.

7.2 Drying Methods

The primary advantage of drying foods and beverages is that they can be preserved for a longer duration. At the same time, the food industry has utilized drying techniques to make convenient ready-to-reconstitute food mixes, which in itself has now become a different sector representing 'instant food.' These foods are better than conventional dried food items as they serve as functional foods with added health benefits and sensory acceptance. There are many drying techniques used to produce instant foods such as spray drying, freeze drying, and foam-mat drying (Yoha et al., 2023). These techniques are product-specific, increase productivity, and achieve better control of the process to enhance product quality. The rehydration ability is also of utmost importance when selecting a drying method for a specific product.

7.2.1 Spray Drying

Spray drying is among the most extensively used drying methods in the food industry for the development of F&V juice powders, milk powder, etc. The process deals with the atomization of a solution, comprising a mixture of components required to make a desirable product, into small droplets, followed by the fast evaporation by hot air at a specific inlet–outlet temperature to produce solid dried powder. The inlet –outlet temperature, pressure, and air flow rate can be optimized based on the type of food product as well as the probiotic strain used. One of the major advantages associated with this drying technique is that it allows the production of nano- and microcapsules, thereby preventing the degradation of associated nutrients, flavors, oils, and other desirable components. Additionally, the simple, fast, and economical operation of spray dryer, with a flexibility to use different coating substances, makes it a desirable technique with high efficiency to produce encapsulated powder for the commercial food sector (Misra et al., 2022).

Furthermore, the literature encompasses an extensive data on the use of spray drying for making dairy, fruits, vegetables, cereals, and legume-based beverage powders. However, the incorporation of synbiotic systems in powder-based matrices is still emerging. Few researchers have explored the potential of different food matrices as primary prebiotic sources that, when clubbed with an appropriate probiotic microorganism, yield a synbiotic mixture. Hence, spray drying allows generation of a novel functional food in dried form with enhanced shelf life and related health benefits.

7.2.2 Freeze Drying

Freeze drying, also termed lyophilization, is a common technique used to produce high-quality powders in the food industry. This method is used primarily for foods that contain heat-sensitive compounds or components that are prone to oxidation as the process involves low temperatures and a high vacuum atmosphere. It involves dehydration of the product via sublimation, followed by dehydration by desorption. This combination produces microcapsules with protected core material with sustained biological and physicochemical properties for longer times. Hence, the shelf life of freeze-dried powders is generally longer than that of other techniques. Freeze drying also maintains the stability of microencapsulated compounds, especially probiotic microbes. That is why the application of this technique finds its major contribution in the development of various plant-based probiotic or synbiotic food powders. However, the inadequate rates of freezing during the encapsulation of probiotics may sometimes lead to the formation of crystals, which have a detrimental effect on the cell membrane of probiotics. In this way, a compromised viable count of probiotics can be observed in the developed powder. Nonetheless, several combination methods are used to overcome such challenges. For example, freeze drying is often combined with different encapsulation methods, such as drying capsules made via ionic gelation, which improves the ability of the capsules to protect from external environment, allowing a maintained cell survivability with a longer storage time. Freeze drying of foods allows the attainment of required phytochemical concentrations in the encapsulated powder with probiotics, without altering their biological activity and even protecting them from the harsh gastric conditions.

7.2.3 Foam-Mat Drying

Foam mat drying is a cost-effective alternate to spray, freeze, and drum drying to produce high-quality food powders. In the process, solid or liquid-based food products are transformed into stable foam via foaming agents, which are whipped along with foam stabilizers. This mixture is then dehydrated using various airflow and thermal methods. The desirable instant powder with superior quality can be attained by carefully selecting the appropriate foaming agents, foaming methods, and foam stabilizers, along with the time taken for foaming, a suitable drying method, and adequate drying temperatures.

This process can be used for large-scale manufacture because of the retention of nutritional quality, easy and faster reconstitution properties, as well as a faster drying rate at a lower temperature. Furthermore, it is an inexpensive method compared to both freezing and spray drying, making it more accessible to the smaller food manufacturers in the market. In literature, an extensive work has been investigated using this technique or a variety of food powders made using dairy, fruits, vegetables, and cereal-based matrix.

7.3 Advantages of Powdered Synbiotic Foods

Apart from being a convenience product that is easy to make, store, and consume, instant food powders with the essence of synbiotics in them exert several other noteworthy advantages for humankind.

7.3.1 Longer Shelf Life

The phenomenon of drying is known for extending the shelf life of food products via the removal of moisture from them. Additionally, the lower water activity is related to the lesser chances of microbial spoilage; hence, the quality of food products during storage is also maintained. Among the different food groups, pro/synbiotic-based dairy food powders like milk powders, ice cream mixes, and yogurt powders are convenient to consumers with a shelf life of more than six months in comparison to their raw counterparts. Nonetheless, the non-dairy substitutes are superior, with an even higher shelf life of more than one year. Hence, the limitations associated with dairy-based products can easily be overcome by using non-dairy synbiotic alternatives. The shelf life of the synbiotic food powder might be affected by the drying treatment it has undergone. This is because the extreme temperature exposure to a food product alters the functionality of the probiotic microbes or the prebiotic components, which are desirable in the final product. Hence, the smart and appropriate use of drying technique with respect to the probiotic and prebiotic used along with the carrier vehicle is of utmost importance while considering the longer shelf life of a product.

7.3.2 Specialty Food

Synbiotic instant food powders like ready-to-reconstitute drinks and instant mixes, when made using certain special ingredients, hold the potential to serve specific sectors of society. As previously discussed, synbiotic non-dairy substitutes in the form of instant food products can be used to fulfil the needs of those suffering from milk allergies, cholesterol-related issues, those who cannot tolerate lactose, or those who in general follow veganism or are vegetarian. Moreover, the growing consumer demand for the non-dairy food industry as a result of general awareness and health consciousness has led to the development of pro/synbiotic-based ready-to-eat functional foods. Such products also provide variety in limited food options. Hence, the integration of probiotics and prebiotics into the plant-based food items followed by an additional step to dehydration creates more space for innovation in terms of health benefits to humankind via instant food items that simplify the lifestyle.

7.3.3 Health Benefits

The consumers in today's world are well-versed in the health benefits associated with functional foods. Instant food powders with added nutrition via incorporation of probiotics, prebiotics, or synbiotics are the trending food options, not because of the accessibility but because of the advantages they exert on human health (Bauer-Estrada et al., 2023). Synbiotics allow the synergistic functioning of probiotics and prebiotics, which enhance the human gut strength by controlling intestinal infections, inhibiting pathogens, and increasing the growth of good bacteria. Moreover, such products are comparatively superior in the nutrition profile to a non-pro/synbiotic product. Additionally, the type of delivery matrix (dairy or non-dairy) also adds to the nutritional content of the final product. Though, the high-temperature drying often raises the question of preservation of desirable functionality of probiotics in dried synbiotic food powders. In this regard, various studies have investigated the optimization of the most suitable drying conditions, packaging materials as well as inoculum concentration for the survival probiotic under the simulated gastric environment. Numerous studies have also projected the benefits associated with the use of microencapsulation, specially for synbiotic instant food powders, to protect the probiotic microorganisms from harsh processing conditions. Such products can be employed at the commercial level with desirable-level probiotics ($>10^7$ CFU/mL).

7.4 Summary

Instant food products with the added benefits of synbiotics allow convenience and simplify the day-to-day lifestyle.

Ready-to-consume synbiotic food powders in the form of juices, beverages, milk, and ice cream mixes exhibit a scope for innovation in the food sector, specially the non-dairy industry.

When the delivery matrix is replaced by a non-dairy food group, such a product acquires an additional sector in the food industry, which includes lactose-intolerant people or those who do not consume dairy products in general.

Synbiotic dairy as well as non-dairy instant food powders have desirable recognition by customers and are growing well in the functional food market.

The idea of drying such products allows the development of a product with a longer shelf life, conserved nutrients, beneficial effects, and expediency to the consumers.

7.5 Multiple Choice Questions

1. Which one of the following does not belong to the advantages of synbiotic instant food powders?
   a. Extended shelf life
   b. Easy bulk handling
   c. Flexibility in matrix selection
   d. Limited products

2. Synbiotic instant food powder can be made using ___.
   a. dairy items
   b. fruits and vegetables
   c. cereals and legumes
   d. All of these

3. Food waste can be used as a delivery matrix for synbiotic powder because _____.
   a. they are a rich source of secondary metabolites
   b. they lack water-holding, and gel formation abilities
   c. Both a and b
   d. None of these

4. _____ is the most widely used drying method in the food industry used for the development of F&V juice powders, milk powder, etc.
   a. Freeze drying
   b. Spray drying
   c. Foam mat drying
   d. Drum drying

5. ____ method is employed for foods that contain heat-sensitive compounds or components that are prone to oxidation.
   a. Freeze drying
   b. Spray drying

    c. Foam mat drying

    d. Drum drying

6. In foam mat drying of instant synbiotic foods, high-quality food powder attributes can be obtained by the selection of ____.

    a. appropriate foaming method, foaming agents, and foam stabilizers

    b. time taken for foaming

    c. suitable drying method and temperature

    d. All of these

7. What are the advantages of synbiotic instant food powder?

    a. Healthy

    b. Shelf stable for a longer duration

    c. Option for people with different needs

    d. All of these

8. ___ is used specially for synbiotic instant food powders to protect the probiotic microorganisms from harsh processing conditions.

    a. Preservatives

    b. Microencapsulation

    c. Chilling

    d. Syneresis

9. Factors contributing to the high nutrition of synbiotic instant foods are ____.

    a. type of delivery matrix

    b. probiotics and prebiotics

    c. Both a and b

    d. None of the above

10. Freeze drying can be harmful to synbiotic food powder because of _____.

    a. inadequate rates of freezing during the encapsulation of probiotics

    b. formation of crystals that have a detrimental effect on the cell membrane of probiotics

    c. Both a and b

    d. None of the above

7.6 Short Answer Type Questions

Q1. Write a short note on the processing of synbiotic dairy-based instant powders.

Q2. Discuss in short the potential of various non-dairy synbiotic food products.

Q3. List down the advantages associated with powdered synbiotic foods and elaborate more on either aspect.

Q4. Elaborate on any one drying technique used for the development of powdered synbiotic foods.

7.7 Descriptive Questions

Q1. Explain in detail the difference between dairy and non-dairy synbiotic food industries based on your perception of one being better than the other.

Q2. With a suitable example, discuss the major benefits associated with the consumption of instant synbiotic food items.

Q3. Discuss in detail the different techniques for drying synbiotic food products at a commercial level. Comment on the challenges faced by this sector.

7.8 Answers for MCQs

Q1	Q2	Q3	Q4	Q5	Q6	Q7	Q8	Q9	Q10
d	d	a	b	a	a	d	b	c	c

References

Bauer-Estrada, K., Sandoval-Cuellar, C., Rojas-Muñoz, Y., & Quintanilla-Carvajal, M. X. (2023). The modulatory effect of encapsulated bioactives and probiotics on gut microbiota: Improving health status through functional food. *Food and Function, 14*(1), 32–55. 10.1039/d2fo02723b

Chaturvedi, S., Khartad, A., & Chakraborty, S. (2021). The potential of non-dairy synbiotic instant beverage powder: Review on a new generation of healthy ready-to-reconstitute drinks. *Food Bioscience*, *42*, 101195. 10.1016/j.fbio.2021.101195

Gervasi, C. A., Pellizzeri, V., Lo Vecchio, G., Vadalà, R., Foti, F., Tardugno, R., Cicero, N., & Gervasi, T. (2022). From by-product to functional food: The survival of *L. casei shirota*, *L. casei immunitas* and *L. acidophilus johnsonii*, during spray drying in orange juice using a maltodextrin/pectin mixture as carrier. *Natural Product Research*, *36*(24), 6393–6400. 10.1080/14786419.2022.2032049

Kalita, D., Saikia, S., Gautam, G., Mukhopadhyay, R., & Mahanta, C. L. (2018). Characteristics of synbiotic spray dried powder of litchi juice with *Lactobacillus plantarum* and different carrier materials. *LWT*, *87*, 351–360. 10.1016/j.lwt.2017.08.092

Rashidinejad, A., Bahrami, A., Rehman, A., Rezaei, A., Babazadeh, A., Singh, H., & Jafari, S. M. (2022). Co-encapsulation of probiotics with prebiotics and their application in functional/synbiotic dairy products. *Critical Reviews in Food Science and Nutrition*, *62*(9), 2470–2494. 10.1080/10408398.2020.1854169

Upadhyay, R., & J, P. D. (2021). Physicochemical analysis, microbial survivability, and shelf life study of spray-dried synbiotic guava juice powder. *Journal of Food Processing and Preservation*, *45*(2), e15103. 10.1111/jfpp.15103

Suggested Readings

Misra, S., Pandey, P., Dalbhagat, C. G., & Mishra, H. N. (2022). Emerging technologies and coating materials for improved probiotication in food products: A review. *Food and Bioprocess Technology*, *15*(5), 998–1039. 10.1007/s11947-021-02753-5

Yoha, K., Moses, J. A., & Anandharamakrishnan, C. (2023). Effect of different drying methods on the functional properties of probiotics encapsulated using prebiotic substances. *Applied Microbiology and Biotechnology*, *107*(5–6), 1575–1588. 10.1007/s00253-023-12398-3

8

SCREENING AND EVALUATION OF SYNBIOTICS

8.1 Screening Methods

Screening of isolated microorganisms and different probiotic ranges is essential for classifying them based on their target functions, technological applications, and possessed properties. The screening process usually consists of a comprehensive approach with multiple steps (Figure 8.1).

For the development of a synbiotic product, the first requirement is screening of various factors, which is described in detail below:

8.1.1 Selection of Probiotics

As per the guidelines issued by the Food and Agriculture Organization (FAO), World Health Organization (WHO), and the European Food Safety Authority (EFSA), probiotic microbial strains must be safe and functional and have a technological utility. In general, probiotic properties such as the origin of the strain, its interaction and effect in the presence of a disease-causing agent, and the antibiotic resistance activity contribute to the overall safety of the strain (Table 8.1). One of the aspects that determines the functional property of probiotics is their ability to endure the harsh GI system and exhibition of their immunoregulation activity (Geng et al., 2023). Moreover, the type of delivery material is necessary because it can directly alter the viability of the specific probiotic. Therefore, probiotics must also satisfy the specific criteria as per the manufacturing or formulation of that carrier matrix. Probiotics should be capable of surviving and, at the same time, maintaining their quality till the storage and delivery step. Nonetheless, probiotics must also exhibit the required health benefits to enhance the functionality of the final marketed product.

DOI: 10.1201/9781003304104-8

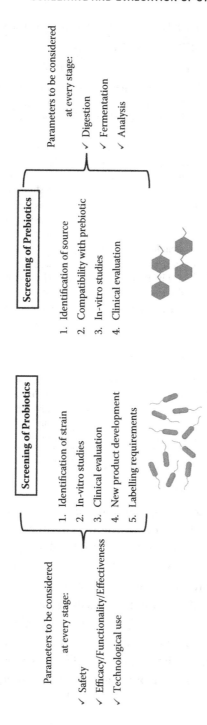

Figure 8.1 Screening parameters for probiotics and prebiotics for the development of synbiotic food products.

Table 8.1 Criteria for the Selection of Probiotics and Their Required Features

PROBIOTIC'S SELECTION CRITERIA		
FUNCTIONALITY	SAFETY	TECHNOLOGICAL USE
• Ability to exhibit competitive exclusion toward microbial species and pathogens inhabiting in intestine.	• Must be originated from animal, human, or fermented food items.	• Must produce a larger biomass yield and should be easy to produce.
• Ability to thrive, remain viable, and retain their metabolic processes.	• Must have previous record of safety and or proof for reliability.	• Must be tested for survival and stability throughout production, processing, and distribution steps.
• Ability to withstand gastric enzymes, gut's acid, and intestinal bile salts.	• Must be diagnosed with accurate genotype and phenotype traits identification.	• Must exhibit a desirable survival rate in the final product.
• Ability to exhibit antimicrobial properties.	• Must not be related to any infection or disease.	• Must not alter the desired characteristics of the final product.
• Ability to exert inhibitory effect toward pathogens like *Listeria monocytogenes*, *Helicobacter pylori*, and *Clostridium difficile*.	• Must not have any toxic profile.	• Must maintain stability of gene.
• Ability to adhere to gut epithelial cells and must survive for the longest duration possible.	• Must not to degrade the bile salts.	• Must not produce undesirable by-products that are capable of altering the characteristics of a product.

8.1.2 Selection Criteria for Prebiotics

Prebiotics are the known supplement for probiotics, and the combination of the two also leads to the development of a synbiotic system. Prebiotics allow the growth of gut microbiota, which in turn initiates the fermentation process. They also lower the pH of stomach and allow better retention of osmotic water in the gut. However, consumption of prebiotics in excess causes the chances of diarrhea, erosion of intestinal lining, and other gut-related problems.

When the prebiotics are taken in an appropriate amount, they exhibit desired functions without any side effects. Prebiotics are, in general, non-allergic until someone has a specific allergy to some specific prebiotic (Grimoud et al., 2010). Prebiotics have various properties, as listed below, that must be considered while selecting it for any synbiotic product.

a. Prebiotics should not be partially or fully digestible in the GI tract.
b. Prebiotics must be fermented by beneficial gut bacteria along with probiotic microorganisms.
c. Prebiotics must exhibit a positive health impact via metabolite production as a result of fermentation, increase stool bulk, growth, and colonization of beneficial gut microbes, enhanced gut health, and improved immune function.
d. Prebiotics must selectively allow the growth and activity of the beneficial microorganisms and resist the growth of other pathogenic ones.
e. Prebiotics must maintain their stability and endure activity during food processing in the human gut and other expected environments.

8.2 Evaluation Techniques

8.2.1 Evaluation Techniques for Probiotic Properties

8.2.1.1 Resistance to Simulated Gastric and Intestinal Fluids Resistance of probiotics to harsh gastric and intestinal fluids is a required property. To evaluate this property, a typical protocol has been presented here. Initially, a simulated GI juice (HCl: 0.1 mol L^{-1}) at a pH between 1.5 (during fasting) and 3 (during normal diet) is made. Similarly, simulated intestinal fluid (NaHCO$_3$: 0.1 mol L^{-1}) is made using pancreatin enzymes and bovine bile salts at appropriate quantities and at a final pH of 8.0. The aim of carrying out this gastric simulation is to determine the probiotic's survivability and concentration via fermented food in the stomach, duodenum, and ileum under the presence of gastric enzymes and changing pH of the GI tract. About 1.0 g of food sample is added in 9.0 mL of saline solution, and the initial pH is adjusted to 6.9 based on the acidity of the mouth, and it is left to stay for 2 min. In the next step or esophagus-stomach stage, the earlier prepared simulated gastric juice was used to adjust the pH of the solution with increasing time (minutes), as follows: 5.5/10, 4.6/10, 3.8/10, 2.8/20, 2.3/20, and 2.0/20 at 37°C. Meanwhile, the solution must be rotated at 130 rpm to replicate the peristaltic movement in the stomach. In the next step, referring to the duodenum, the simulated bile juice is added to obtain a pH of 5 along with bile enzymes and pancreatin, and incubated for 20 min at 45 rpm at 37°C. Similarly, in the next step indicating ileum, the pH is increased to 6.5 and subjected to incubation at 37°C for 90 min (90 rpm). In the end, a desirable sample quantity can be analyzed to enumerate the probiotic cells surviving the simulated mouth-to-gut passage.

The growth of probiotics at every step of the simulation can be used to determine the initial inoculation concentration of the probiotic in the final product to achieve the recommended range of probiotic counts till the storage of the product.

8.2.1.2 Antibacterial Ability or Growth Inhibition of Pathogenic Strains Probiotics are known to produce antimicrobial substances such as organic acids, bacteriocins, and enzymes, which inhibit the growth of pathogenic bacteria and promote good intestinal health. The inhibition of the growth of pathogenic strains by probiotics must be evaluated to determine its efficacy. For this purpose, 'agar spot test' is used wherein probiotic bacteria having a survival count of 1×10^8 CFU/mL are spotted on an MRS agar plate and incubated anaerobically at 37°C for 48 h. On the other hand, pathogenic strains cultured in 5 mL of soft agar are prepared and poured onto MRS agar with probiotic spots. The plates are then subjected to incubation at the conditions specific to the pathogen. Later, based on the diameter of inhibition in millimeters, the inference can be made. If the diameter is >50 mm, the efficacy of the probiotic in inhibiting the growth of probiotics is good, while no or small diameter indicates the inability of probiotics to inhibit the pathogen.

8.2.1.3 Utilization of Probiotics by Prebiotic Extracts To evaluate whether the prebiotic has been utilized as the source of carbon for lactic fermentation, by the probiotic used, a stepwise method is used (Huynh et al., 2021). Firstly, the probiotic bacteria are cultured in MRS broth followed by centrifugation and two consecutive washing. The final culture in a modified-MRS medium (m-MRS) is then plated on m-MRS agar plates with 2% prebiotic extract and bromocresol purple indicator. The ability of the probiotic to utilize prebiotic extract can then be observed from the discoloration of the surrounding area of the agar plate from violet to yellow. For comparative analysis, negative and positive controls can be used with probiotics cultured on m-MRS plates with no prebiotic/ glucose and with glucose, respectively. In this way, the efficacy of probiotics to utilize prebiotics as a potential feed can be verified.

8.2.1.4 Growth Stimulation of Probiotics by Prebiotic Extracts The growth of probiotic strains in the presence of prebiotic extract can be assessed by measuring the rise in absorption at 600 nm after 0, 2, 4, 6, and 24 h using a spectrophotometer. The change in pH can also be observed simultaneously as an indirect indicator of the growth as well as the breakdown of prebiotic components in the extract.

8.2.1.5 Adhesion Ability Adhesion ability is a classical selection condition for potential probiotic bacteria as the eventual colonization will allow the exhibition of numerous benefits such as immunomo- dulatory effects, gut barrier properties, metabolism-related functions, production of antimicrobial substances, and reduction of pathogenic adhesion via competitive exclusion. Moreover, prebiotics also prevent the adhesion of toxins and pathogens to epithelial receptors. Hence, the evaluation of the adhesion ability of probiotics is necessary and can be done via the following methods:

a. Cell surface hydrophobicity assay

   The hydrophobicity exhibited by the outermost surface of microorganisms is related to its adhesion to host tissue. This property is essential and provides a competitive advantage to the probiotics in the maintenance of the human GI tract The cell surface hydrophobicity of the probiotic with prebiotic(s) is determined by collecting the thoroughly washed probiotic bacteria in sterile phosphate-buffered saline with a final absorbance of 0.80 at 600 nm. Thereafter, 3 mL of the bacterial suspension and 1 mL of xylene are mixed and incubated at room temperature. The cell surface hydropho- bicity can be then calculated as per equation (8.1):

$$\text{Cell surface hydrophobicity}(\%)$$
$$= [(A_0 - A_t)/A_0] \times 100\% \qquad (8.1)$$

where A_0 and A_t are the absorbance values of the probiotic before and after xylene treatment, respectively. The higher the surface hydrophobicity, the better the aggregation and coaggregation properties.

b. Auto-aggregation assay

Aggregation ability is the property of bacterial cells to adhere, survive, and persist in the human GI tract. Auto-aggregation ability can be calculated by allowing the probiotic culture under xylene treatment for a longer period.

$$\text{Auto} - \text{aggregation ability}(\%)$$
$$= (1 - A_t / A_0) \times 100\% \qquad (8.2)$$

where A_t and A_0 are the absorbance values taken at 600 nm at time t (h) and 0 h, respectively.

c. Coaggregation assay

Coaggregation is considered an essential property in eliminating pathogens from the GI tract by serving as a barrier and preventing colonization by pathogenic bacteria. Coaggregation with a potential pathogenic microbe in very close proximity allows the probiotic to produce antimicrobial substances, which may inhibit the growth of pathogenic strains in the GI tract. Auto-aggregation and coaggregation are significant phenomena used in forming biofilm to protect the host from pathogenic colonization. For the evaluation, equal volumes of probiotic strain and pathogenic bacterium are mixed, incubated at 37°C for 2 h and later the absorbance is measured at 600 nm.

$$\text{Co} - \text{aggregation ability}(\%)$$
$$= [1 - A_{mix}/(A_{probiotic} + A_{pathogen})/2] \times 100\% \qquad (8.3)$$

d. Adhesion to Caco-2 cells

Unlike auto-aggregation, coaggregation, and hydrophobicity, determining the adherence ability of probiotics by creating human enterocyte models like Caco-2 or HT-29 cell cultures is slightly time-consuming and expensive. For this assay, Caco-2 cells are added to a 12-well plate, followed by incubation at 37°C in a saturated atmosphere with 5% CO_2

and 95% air until a monolayer is attained. Later, an antibiotic and fetal bovine serum-free probiotic suspension are added to the wells and further incubated for 1 h at 37°C. Then 1 mL of 1% Triton X-100 is added, and the mixture is stirred to detach the bacterial cells from Caco-2 cell monolayers. Serial dilution of the suspension is then plated onto MRS plates and incubated to determine the viable bacterial cell number.

8.2.1.6 Antibiotic Susceptibility Probiotics are known to produce antibacterial substances that antagonize pathogenic microbes, competing for adhesion sites and thereby protecting the gut from pathogens and exhibiting immune regulation. The synbiotics can be evaluated for their antibiotic susceptibility by disk method, where antibiotic paper disks are placed on agar plates containing synbiotics and incubated for 24 h at 37°C. The diameter of the inhibition zone is then measured, and the susceptibility of the synbiotics to different antibiotics can be determined using the standard literature on antibacterial susceptibility. The susceptibility of a probiotic to antibiotics is necessary because studies have shown that the antibiotic-resistance genes in probiotics may spread to other pathogenic bacteria, causing compromised health (Muganga et al., 2015).

8.2.1.7 Determination of SCFAs Production Short-chain fatty acids (SCFAs) exhibit various health benefits on the human body. They can be produced through the fermentation of prebiotics by probiotics along with the internal gut microbiome. To determine the types and amounts of SCFAs produced during the fermentation process, the 48 h incubated samples (free from any insoluble material) are collected and filtered through membrane filters. The growth media supernatant is used to quantify the SCFA productivity via reverse phase high-performance liquid chromatography (RP-HPLC) equipped with a C18 column.

8.2.2 Evaluation Techniques for Prebiotic Properties

8.2.2.1 Prebiotic Activity Score The prebiotic activity score (PAS) indicates the ability of a prebiotic(s) or the potential oligosaccharides to enhance the growth of probiotics in comparison to standard sugar 'glucose' and to inhibit the growth of pathogenic microbes. A positive and high prebiotic score denotes that the probiotics grow better in the presence of prebiotics than glucose and do not allow the growth of pathogenic bacteria. Similarly, a low and negative prebiotic score indicates the growth of beneficial probiotic bacteria with prebiotics lower than with glucose and/or increased pathogen growth. The PAS is assessed using serial dilutions and plating techniques and calculated using the following equation:

Prebiotic Activity Score (PAS)=

$$\frac{\begin{array}{l}\text{(growth of probiotic in presence of prebiotic at 24 h}\\ - \text{ growth of probiotic in presence of prebiotic at 0 h)}\end{array}}{\begin{array}{l}\text{(growth of probiotic in presence of prebiotic at 24 h}\\ - \text{ growth of probiotic in presence of prebiotic at 0 h)}\end{array}}$$

$$\frac{\begin{array}{l}\text{(growth of pathogen in presence of prebiotic at 24 h}\\ - \text{ growth of pathogen in presence of prebiotic at 0 h)}\end{array}}{\begin{array}{l}\text{(growth of pathogen in presence of prebiotic at 24 h}\\ - \text{ growth of pathogen in presence of prebiotic at 0 h)}\end{array}}$$

8.2.2.2 Probiotic Enzyme Activity Induced by Prebiotic Component(s) The ability of probiotics such as those belonging to *Lactobacillus* sp. in exerting enzymatic activity is considered an important criterion in the selection and utilization of prebiotics. This is because the consumption of prebiotics is associated with proteolysis, which in turn maintains the probiotic biomass growth. Hence, a prebiotic must show the highest activity of protease derived from probiotic bacteria like *Lactobacillus*. For this, the probiotic is cultivated in modified MRS broth supplemented with sample prebiotic and standard glucose (control) and subjected to incubation at 37°C. Later, the cells are removed and centrifuged to obtain the cell-free supernatants. The protease activity of this supernatant is assayed at 40°C in 100 mmol/L Tris-HCl buffer (pH 9.0). The data from this test can be used to observe whether the *Lactobacillus* sp. exhibits higher or lower production levels of protease enzymes, which impact the protein digestibility of the prebiotic source as compared to those in the glucose control.

8.3 Summary

- The knowledge of the required parameters for the selection, screening, and evaluation of probiotics and prebiotics is essential for the delivery of a synbiotic product commercially.
- It could create ethical and legal issues with compromised health at the same time.
- Following the given set of criteria, the selection of prebiotics and probiotics becomes crucial.
- While functionality, safety, and the intended technological use are the required criteria for the selection of a probiotic bacteria, for prebiotics, the substance must be compatible with the selected probiotic with a positive impact on human health.
- The evaluation techniques assessed for probiotics include properties such as resistance toward simulated gastric and intestinal fluids, antibiotic susceptibility, and growth stimulation by prebiotic extracts.
- Similarly, a good prebiotic activity score and probiotic enzyme activity induced by prebiotic components are evaluated while screening a prebiotic for a synbiotic food system.

8.4 Multiple Choice Questions

1. Which among the following are the criteria for the selection of probiotics?
   a. Functionality
   b. Safety
   c. Technological use
   d. All of the above
2. The ability of probiotics to exhibit competitive exclusion toward microbial species and pathogens inhabiting in intestine comes under the criteria of ___.
   a. functionality
   b. safety
   c. technological use
   d. All of the above

3. Which among the following is a required feature for the safety of probiotics?
   a. Ability to exhibit antimicrobial properties.
   b. Must not be related to any infection or disease.
   c. Must not alter the desired characteristics of the final product.
   d. Ability to exert inhibitory effect toward pathogens like *Listeria monocytogenes.*

4. Which among the following is NOT correct for prebiotics?
   a. Prebiotics should not be digestible or partially digestible in the GI tract.
   b. Prebiotics must not be fermented by beneficial gut bacteria along with probiotic microorganisms.
   c. Prebiotics must selectively allow the growth and activity of the beneficial microorganisms and resist the growth of other pathogenic ones.
   d. Prebiotics must maintain their stability and endure activity during food processing in the human gut and other expected environments.

5. The _____ nature of the outermost surface of microorganisms is related to the attachment of bacteria to host tissue.
   a. hydrophobic
   b. hydrophilic
   c. lipophobic
   d. lipophilic

6. _____ is the property of bacterial cells to adhere, survive, and persist in the human GI tract.
   a. Antibacterial activity
   b. Prebiotic activity
   c. Tolerance ability
   d. Aggregation ability

7. A positive and high prebiotic score denotes _____.
   a. probiotics are growing better in the presence of prebiotics
   b. probiotics are growing better in the presence of glucose
   c. prebiotics do not allow the growth of pathogenic bacteria
   d. Both a and c

8. The growth media supernatant is used to quantify the SCFA productivity via _____.

a. reverse phase high-performance liquid chromatography
    b. gas chromatography
    c. atomic absorption spectroscopy
    d. thin layer chromatography
9. The typical pH of the gut during fasting is ____.
    a. 1
    b. 1.5
    c. 2
    d. 2.5
10. Antimicrobial substances produced by probiotics include _____.
    a. organic acids
    b. bacteriocins
    c. enzymes
    d. All of the above

8.5 Short Answer Type Questions

Q1. Discuss the criteria of functionality of probiotics for their selection in a synbiotic system.
Q2. Enlist the various properties considered while selecting prebiotics for a synbiotic product.
Q3. Enlist the various evaluation techniques for probiotics and prebiotics to analyze a synbiotic product.
Q4. Write a short note on 'determination of SCFAs production in the synbiotic system.'
Q5. Write a short note on the 'prebiotic activity score.'

8.6 Descriptive Questions

Q1. Discuss in detail the three major criteria for the probiotic screening and their required features to be followed.
Q2. Enlist the different methods to evaluate the adhesion ability as a potential probiotic selection criterion and discuss any two in detail.
Q3. Differentiate between antibiotic susceptibility and antibacterial ability as the probiotic screening methods.

Q4. Enlist the different methods to evaluate the prebiotic properties and discuss in detail.

Q5. Differentiate between 'utilization of probiotics by prebiotic extracts' and 'growth stimulation of probiotics by prebiotic extracts' as the probiotic screening methods.

8.7 Answers for MCQs

Q1	Q2	Q3	Q4	Q5	Q6	Q7	Q8	Q9	Q10
d	a	b	b	a	d	d	a	b	d

References

Geng, S., Zhang, T., Gao, J., Li, X., Chitrakar, B., Mao, K., & Sang, Y. (2023). In vitro screening of synbiotics composed of *Lactobacillus paracasei* VL8 and various prebiotics and mechanism to inhibits the growth of *Salmonella typhimurium. LWT, 180,* 114666. 10.1016/j.lwt.2023.114666

Grimoud, J., Durand, H., de Souza, S., Monsan, P., Ouarné, F., Theodorou, V., & Roques, C. (2010). In vitro screening of probiotics and synbiotics according to anti-inflammatory and anti-proliferative effects. *International Journal of Food Microbiology, 144*(1), 42–50. 10.1016/j.ijfoodmicro.2010.09.007

Muganga, L., Liu, X., Tian, F., Zhao, J., Zhang, H., & Chen, W. (2015). Screening for lactic acid bacteria based on antihyperglycaemic and probiotic potential and application in synbiotic set yoghurt. *Journal of Functional Foods, 16,* 125–136. 10.1016/j.jff.2015.04.030

Suggested Readings

Huynh, T. G., Vu, H. H., Phan, T. C. T., Pham, T. T. N., & Vu, N. U. (2021). Screening utilization of different natural prebiotic extracts by probiotic Lactobacillus sp. for development of synbiotic for aquaculture uses. *Can Tho University Journal of Science, 13,* 96–105. 10.22144/ctu.jen.2021.022

9

MARKET PROFILE OF
SYNBIOTIC FOODS

9.1 Global Market for Synbiotic Products

The trends with respect to food choices are changing continuously. In today's world, consumers are more health conscious, understand the relationship between diet and health, and look for better and healthier food options. In this view, functional foods, especially those containing probiotic microorganisms or prebiotic components, have additional health benefits and offer a variety in the food industry. Recent successful studies at the lab scale as well as on the industrial level have shown an upsurge in the screening of different probiotics, prebiotics, and synbiotic formulations to deliver them to humans. These probiotics and synbiotic food products have also witnessed an increased commercial interest recently, with increasing annual growth and development of the functional food sector. The health benefits associated with the probiotic bacteria belonging to the *Lactobacillus* family have primarily been used in dairy foods, creating a new niche for healthy, functional foods. Many dairy food companies have increased their market by adding probiotic and synbiotic-based dairy products because of the overall enhanced nutrition and quality of such products, which ultimately result in improved sales. There are more than 100 probiotic products available globally, including milk, beverages, juices, yogurt, buttermilk, powdered milk, chocolates, ice creams, and frozen desserts (Arora, Baldi, 2015). These dairy and non-dairy–based synbiotic products fall under functional food and beverages, contributing to the major global share, followed by dietary supplements, pharmaceutical products, and animal feed, as depicted in Figure 9.1.

In the global market, the United States, Europe, Japan, India, and Australia are the major countries contributing to the production and consumption of synbiotic foods, with approximately 3% of the total

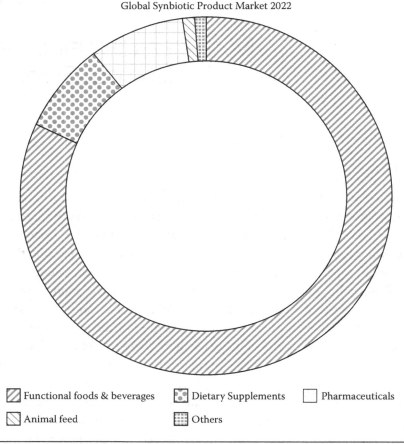

Figure 9.1   Global synbiotic market distribution (product specific) in the year 2022.

food market (Figure 9.2). The consumer acceptance and consumption of different synbiotic and probiotic products differ significantly in various countries, as per their food preferences and cultural beliefs (Dixit et al., 2016). For example, in Europe, the northern states prefer fermented dairy products because of their tradition. Hence, the consumption of fermented dairy products with probiotics and synbiotics has shown maximum growth in this sector. Among other countries, Germany ranks first in the probiotic products market, followed by the United Kingdom and France, respectively. Moreover, North America is also one of the largest probiotic hubs with a huge market size. Still, the consumption of probiotic and synbiotic products in the United States lies more in the category of medical, dietary, and

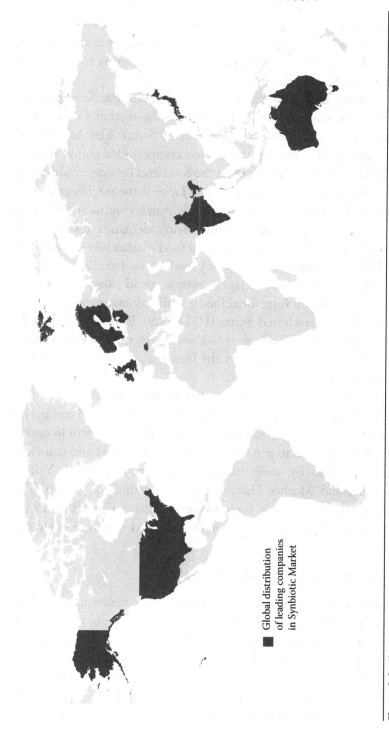

Figure 9.2 Overall global distribution of leading companies in the synbiotic market.

nutritional supplements. America and Canada have over 100 companies for probiotic and synbiotic products, with major market players like Chr. Hansen, Danisco, and Lallemand contributing to about 70% of the total probiotic culture market.

In the latest year's market analysis data, the Asia-Pacific region was estimated to be the largest and fastest-growing market for synbiotic products compared to Europe and the United States. The Asia-Pacific market is projected to grow at the highest compounded annual growth rate (CAGR), with rapid growth in the food and beverage industry in countries such as Japan and India. While Japan is the torchbearer as far as the probiotic sector is concerned in the Asian continent, India, on the other hand, is growing with increasing demands for probiotic functional foods. In Japan, the functional food market was estimated to be over USD 15 billion and has been growing consistently for the last decade. As per recent data, in Japan, people spend close to USD 100 per person annually on functional foods, while others spend comparatively less, like in the United States (USD 67.9), Europe (USD 51.2), and Asia (USD 3.20). This is near to about 5–6% of food expenditure, which is utilized for this aspect by the Japanese, which is very high in comparison to any other country.

Similarly, India also contributes to the overall market of probiotics by occupying <1% of the total global market, with more than 25 companies as established synbiotic product sellers. Nonetheless, a growth in overall market size is expected to rise with the increased interest and launch of more product lines by some leading companies like Amul, Yakult, Nestle India, and Mother Dairy gradually entering the set-up. The probiotic market in India is projected to register a CAGR of 19.8% during 2014–2019, which has shown an increased rate in recent years.

9.2 Market Synopsis of Synbiotic Products

In the last decade, the world probiotic market has flourished exponentially and received significant consumer acceptance owing to the health benefits associated with them. Moreover, the budding branch of probiotics, i.e., prebiotics and synbiotics, has also experienced substantial growth due to the entry of major companies like Nestle (Switzerland). Probiotic and prebiotic-based products from Nestle have enhanced their market and consumption in more than 30 countries globally, especially in the case of infant and dairy products. Nonetheless, the list of countries consuming these products is increasing rapidly due to the popularity of probiotic, prebiotic, and synbiotic-based food products. In 2021, the global synbiotic market was estimated to be USD 817.2 million and is expected to record a profit CAGR of 8.3% in the coming years. This growth is attributed to the rising health awareness among consumers toward nutritional supplements. Moreover, the positive results exerted by symbiotic products in the prevention and treatment of various health issues and diseases and maintenance of overall well-being have also resulted in the surge in the consumption of functional foods, especially, probiotic, prebiotic, and synbiotic-based food products (Cosme et al, 2022). The relevance of synbiotic food in human health, coupled with consumers' rising demand for high-quality food that is safe, convenient, and has a long shelf life, are the major driving growth factors in the functional food market. Additionally, the claims and reports published by various national and international regulatory bodies, which confirm that synbiotic formulations provide safer products with the utmost quality of probiotic and prebiotic components used in them, have led to the adoption of such products even at the industrial levels.

Moreover, integrating functional ingredients, microbial strains, and antioxidants to improve the quality and shelf life of probiotic products are a few of the chief reasons behind the rapid commercialization of synbiotic food products (Corbo et al., 2014). Alongside this, several technical developments, such as nano- and micro-encapsulation, pulse-electric field, high-pressure processing, and ultra-sonication, are showing themselves as potential delivery vehicles for viable probiotic components and prebiotic products. The shifting regime of consumers has led to reduced time for preparing home-cooked balanced food because of the increasing needs of urbanization. Hence, the consumption of quality food is decreasing, which creates the need for quick, ready-to-cook, and ready-to-consume healthy food alternatives. Similarly, convenience in getting food via fast-food outlets and stores is also among the reasons for causing gut-related problems. Thus, the consumption of synbiotics that provide health benefits is the need of the hour, particularly in the maintenance of gut health, improvement in heart-related conditions, and decreased inflammation. All these benefits exerted by synbiotics have led to a higher demand for such products even by various pharmaceutical and nutraceutical companies along with food producers.

9.2.1 Major Form of Synbiotic Product Market Globally

Based on the type of form in which a synbiotic product is sold, the global synbiotic market is segmented into three categories, namely, powder, capsule, and liquid. In 2021, the powder segment reported the maximum revenue share in the global synbiotics market. The major factors that helped to enhance the overall growth of the powder-based products segment include (1) delivery of a high fiber content, which in turn allows better digestion and control blood sugar levels; (2) the treatment of conditions like inflammatory bowel diseases (IBD) and lactose intolerance by exhibiting proper digestion; (3) consumption of digestive enzymes, like amylase, protease, and lipase, which in turn helps in breaking down nutrients and allowing their absorption; and (4) maintenance of the gut microbiome. The powder-based products are majorly utilized by pharmaceutical companies to make pills and capsules. On the other hand, the food industry uses powder to deliver food products as ready-to-reconstitute food items. The major advantages associated with the powder segment compared to liquid or capsule are that it provides stress relief, energy-boosting effects, and immune-boosting via enhanced shelf life and easy uptake.

9.2.2 Application Insights in Global Synbiotic Market

Based on application type, the global synbiotic market is segmented into food and beverages, dietary supplements, pharmaceuticals, animal feeds, and others. Among all, the food and beverage segment accounted for the largest revenue share in the last year owing to their numerous advantages. For example, synbiotic products, when taken as food and beverages, allow direct, better, and maximum nutrient absorption being the primary source. Moreover, the component of probiotics and prebiotics allows additional benefits for gut health and aids in better digestion and weight management. The incline of major companies to develop synbiotic supplements with enhanced nutritional value is increasing because of the demand for functional food for their associated health benefits.

9.2.3 *Major Distribution Channel in the Global Synbiotic Market*

Based on distribution channels, the global synbiotics market is divided into different segments like indirect/B2C (business-to-consumer), online retail, direct/B2B (business-to-business), hyper-markets/supermarkets, retail stores, convenience stores, and others. In 2022, the hypermarket/supermarket segment dominated among others with a market share of around 18%, as the display of a variety of products in supermarkets made buying experience suitable for customers. Moreover, hypermarkets have been budding with time across populated countries such as India, Japan, and China. With the clear visibility of products in different outlets, the easy accessibility to consumers allows for increased revenue growth in this segment. Moreover, growing retailers in the market are also allowing the overall market growth. Rising consumer awareness toward synbiotic products is also among the driving factors for the market revenue growth. In addition, the intervention of new initiatives for synbiotic products with positive health benefits on the human body drives the developers more toward synbiotic products, even for animal feeding and health supplements. Major players in the market, like Danisco and Nestle, are even considering mergers and acquisitions to develop better synbiotic products. Nonetheless, the rising consumer preferences and lifestyle changes are certain other factors driving revenue growth of the synbiotic market in India as well as globally (Saarela et al., 2002). Even companies are now starting to produce synbiotic-based pet food products considering it as one of the leading markets for rapid consumption of synbiotic products. At the same time, the surging demand for dietary supplements by the elderly population has contributed to market growth. These, in turn, increase the use of supermarkets for better reachability and enhanced market growth.

9.3 Summary

The market of synbiotics has shown a rapid increase in recent times.

The popularity of healthy, functional foods has led to consistent growth in this market.

The addition of synbiotics to already existing products as a value-adding ingredient in various huge companies allowed this rapid market upsurge.

India, along with various other developing countries, is still budding in this sector, though the rise seems positive.

The food and beverage sector has contributed the most to the synbiotic market, followed by other sectors like pharmaceutical, medical, dietary, and nutritional supplements.

Synbiotic product is still a niche sector with increasing demand and consumer acceptance.

The individual market contributed by probiotics and prebiotics globally also plays a significant role in the uplifting of synbiotics in general and in the food industry.

9.4 Multiple Choice Questions

1. Functional food and beverages contribute to the major global share of synbiotics, along with _____.
   a. dietary supplements
   b. pharmaceutical products
   c. animal feed
   d. All the above
2. There are more than _____ probiotic products available globally.
   a. 1
   b. 10
   c. 100
   d. 1,000
3. In the global market, synbiotic foods contribute around _____ of the total food market.
   a. 3%
   b. 13%
   c. 30%
   d. 50%
4. In the global market, ____ is/are the major countries contributing to the production and consumption of synbiotic foods.

    a. Pakistan

    b. Iceland

    c. Afghanistan

    d. None of these

5. Globally, companies like _____ contribute to about 70% of the total probiotic culture market.

    a. Danisco

    b. Chr. Hansen

    c. Both a and b

    d. None of the above

6. _____ was estimated to be the largest and fastest growing market for synbiotic products.

    a. Asia-Pacific–based countries

    b. Europe

    c. USA

    d. China

7. The global synbiotic market is expected to grow with a CAGR of _____ in the coming years.

    a. 6.2%

    b. 7.8%

    c. 8.3%

    d. 7.1%

8. _____ for enhancement of quality of probiotic products and shelf life is leading to rapid commercialization of synbiotics.

    a. Removal of functional ingredients, microbial strains, and antioxidants

    b. Incorporation of functional ingredients, microbial strains, and antioxidants

    c. Altering the storage conditions

    d. Selection of suitable packaging material

9. Various technological advancements, such as _____ considered for efficient delivery of viable prebiotic and probiotic products, drive the market growth of synbiotic products.

    a. micro-encapsulation

    b. pulse electric field

    c. high-pressure processing

    d. All of the above

10. Based on the distribution channel, the _____ segment of global synbiotics accounted for a significant revenue share in 2021.
    a. direct
    b. indirect
    c. hypermarkets/supermarkets
    d. None

9.5 Short Answer Type Questions

Q1. Discuss the market synopsis of synbiotics in different Asia-Pacific regions.
Q2. What is the reason behind the enhanced growth in the market of synbiotic food products in certain specific countries?
Q3. What are the major forms of synbiotic product market at the global level?
Q4. Write a short note on the global synbiotic market.
Q5. Discuss in short: major distribution channel in global synbiotic product market.

9.6 Descriptive Questions

Q1. Discuss the market for synbiotic products at the global level.
Q2. Elaborate on how developing countries like India can improve their growth in the synbiotic food market.
Q3. Differentiate between the major forms of synbiotic products marketed globally.

9.7 Answers for MCQs

Q1	Q2	Q3	Q4	Q5	Q6	Q7	Q8	Q9	Q10
d	c	a	d	c	a	c	b	d	c

References

Arora, M., & Baldi, A. (2015). Regulatory categories of probiotics across the globe: A review representing existing and recommended categorization. *Indian Journal of Medical Microbiology, 33*, S2–S10. 10.4103/0255-0857. 150868

Dixit, Y., Wagle, A., & Vakil, B. (2016). Patents in the field of probiotics, prebiotics, synbiotics: A review. *Journal of Food: Microbiology, Safety and Hygiene, 1*(02), 1–13. OMICS Publishing Group. 10.4172/2476-2059. 1000111

Suggested Readings

Cosme, F., Inês, A., & Vilela, A. (2022). Consumer's acceptability and health consciousness of probiotic and prebiotic of non-dairy products. *Food Research International, 151*, 110842. 10.1016/j.foodres.2021.110842

Corbo, M. R., Bevilacqua, A., Petruzzi, L., Casanova, F. P., & Sinigaglia, M. (2014). Functional beverages: The emerging side of functional foods. *Comprehensive Reviews in Food Science and Food Safety, 13*(6), 1192–1206. 10.1111/1541-4337.12109

Saarela, M., Lähteenmäki, L., Crittenden, R., Salminen, S., & Mattila-Sandholm, T. (2002). Gut bacteria and health foods—The European perspective. *International Journal of Food Microbiology, 78*(1–2), 99–117. 10.1016/S0168-1605(02)00235-0

10

COMMERCIAL ASPECTS AND CHALLENGES WITH SYNBIOTICS

10.1 Commercialization of Synbiotics in the Marketplace

In the marketplace, synbiotic products are delivered to consumers in different forms, such as food, beverages, and pharmaceuticals. While probiotic products have conquered the functional food market for a long time, the use of prebiotic and synbiotic products has also been observed in the industry lately. A majority of the products are still sold in the name of probiotics instead of synbiotics. This is because of the unpopularity of the word 'synbiotic product.' Particularly, synbiotics have received considerable consumer attention in recent years despite not being a common food supplement. Moreover, the application of synbiotics in dairy and non-dairy food products has shown great potential in the food industry, with numerous benefits associated with them. Synbiotic-based food products in dairy matrix allow added health benefits via the consumption of good microorganisms and healthy prebiotics, hence serving the dietary requirements of health-conscious consumers. Similarly, in the non-dairy matrix, such a product holds the potential to serve as an option for vegan or lactose-intolerant consumers who demand such products.

10.2 Challenges with Synbiotics

In spite of these market prospects, several industrial and scientific barriers create difficulties for synbiotic-based food products. The major challenges faced by synbiotic products at the market level have been discussed in detail (Figure 10.1).

10.2.1 Physiological Challenges

Any product containing a live and viable microorganism must undergo regulatory channels before reaching the consumers.

 DOI: 10.1201/9781003304104-10

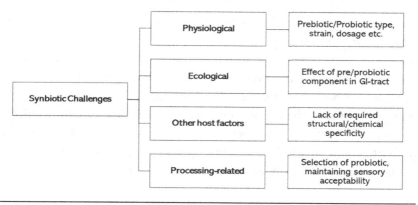

Figure 10.1   Major challenges faced by synbiotic products at the market level.

Different countries follow different rules and regulations for such products. For example, any product containing probiotics must represent a health claim with it. Similarly, synbiotic products, which are a part of probiotics, must also undergo such scrutiny procedures in order to serve the consumers legally. Additionally, the prescribed amount of dosage by regulatory authorities is also recommended when delivering such products to the market. Hence, these are considered the physiological challenges associated with synbiotic products owing to their effects on the host's health.

10.2.1.1 Establishment of Clinical Efficacy   One of the main barriers faced by synbiotic foods is difficulty in indicating the clinical efficacy of the developed product. This means whether the developed synbiotic product possesses desirable evidence(s) required to support a health claim, which varies from country to country. Currently, neither the European Food Safety Authority (EFSA) nor the Food and Drug Administration (FDA) has approved any health claims made for probiotic or synbiotic combinations. Moreover, a clear distinguishing factor for classifying a product as either a pharmaceutical or a food product is lacking. Hence, these factors create doubt in the minds of consumers to trust the product, in turn restricting the market (Kumar et al., 2023).

10.2.1.2 Establishment of Minimum Effective Dose   Determining the minimum effective dose of each component in a synbiotic product is usually expected. For this, the presence of appropriate control samples for synbiotic trials is necessary, but challenging at the same time. Moreover, the selection of standards for the probiotic and prebiotic components is also required to confirm the overall synergistic effect of the product (Arora, Baldi, 2015). The selection of these components for various food products, despite such extensive research, remains challenging because of the wide range of probiotic species and strains, the types of prebiotic sources, the dosage of probiotics and prebiotics, and the specific target outcomes of each component, individually as well as in combination.

10.2.2 Ecological Challenges

A few ecological constraints also influence the ability of probiotic microorganisms and/or prebiotic components to alter the functioning of the GI tract. Several probiotic products consist of microbes from different sources and origins that cannot thrive in gastric conditions. This holds a challenge in the selection of appropriate strains of the probiotic in a product as per its specificity. Similarly, few probiotic microorganisms are not compatible with the prebiotic components or the non-digestible carbohydrates in breaking and consuming it. This is one of the major aspects that determines the suitability of a probiotic in the human gut environment. Similarly, the efficiency of prebiotics alone is also crucial as a selective substrate to probiotic microorganisms. It allows their consumption and conversion to beneficial by-products, which in turn have additional health benefits.

Hence, these ecological challenges, along with physiological and related host factors, need to be taken care of while developing a synbiotic product.

10.2.3 Challenges during Processing

Regulation of the processing of synbiotic products from the beginning of production to the end of the shelf life is needed to preserve consumer acceptance in terms of nutritionally rich and healthy products with viable probiotic counts.

10.2.3.1 Selection of Probiotic Bacteria/Strain The selection of appropriate probiotic microbes is essential such that the sensory appeal and the shelf life of the product are not compromised. The selected probiotics must be compatible with the prebiotic component and withstand the harsh conditions encountered during food processing (high temperatures, acidic pH, and oxidative and osmotic stress levels) as well as the severe gastric environment. Similarly, adhering to safety regulations and specifications related to the microbial strain like pathogenicity, toxicity, infectivity, metabolic activity, and intrinsic factors is equally important for the selection of probiotics (Koirala, Anal, 2021).

10.2.3.2 Effect of Processing Conditions on Synbiotic Formulations Numerous processing factors directly influence the viability of probiotics. These factors include the processing factors (product's treatment temperature, fermentation time and temperature, related incubation condition, cooling and storage temperatures, type of packaging materials, etc.) and food matrix factors (pH, acidity, water activity, oxygen level, salt-sugar concentrations, and presence of enzymes and chemicals like hydrogen peroxide, coloring agents, and artificial flavoring) (Quintero et al., 2022). Apart from this, based on the type of processing applied, the processing factors vary and must be considered accordingly in order to prevent any kind of damage to the probiotic as well as the food material such that the desirable features sustain.

10.2.3.3 Sensory Acceptability Preservation of the sensory quality of developed synbiotic products owing to their exposure to harsh processing conditions is crucial in determining consumer acceptance. Moreover, processing conditions have a direct relationship with product nutritional quality and consumer acceptability. Several studies have successfully developed non-dairy synbiotic beverages with acceptable consumer preferences, such as ready-to-drink ice tea from black and green tea with added prebiotics (GOS, FOS, and inulin), and synbiotic cashew milk using probiotic *B. animalis*. Nonetheless, careful evaluation of the sensory acceptability of the final synbiotic product must be taken care of as process parameters, along with fermentation, cause development as well as loss of certain flavoring compounds, which in turn result in a unique flavor of the product.

10.3 Future Recommendations for Categorization of Synbiotics

Probiotics, prebiotics, and synbiotics have now been proven by many specialists to confer numerous health benefits and can also be used in the treatment of some diseases. Though, the related health claims and labeling are too specific and lack scientific evidence, causing a major concern toward false claims and gimmicks for promotion by some companies. Hence, in this regard, it is recommended that a clear and precise definition and labeling should be followed for probiotic, prebiotic, and synbiotic food and beverage-based products (Pineiro, Stanton, 2007). One of the major reasons for the use of multiple definitions and misleading labels is the lack of universal regulations and adequate legislation across the globe. Different countries tend to follow different rules with respect to the use of live microorganisms in any food item, hence creating complexity in the system. At the same time, there is also a need to report the issues related to the quality, safety, and efficiency aspects of probiotics by defining the limits of its categories. The issues related to diverse categorization can be resolved by considering the following points:

- The adoption of universally accepted standard definitions and rules is the need of the hour for creating a basic background and harmonized regulatory framework of guidelines for symbiotic products.
- Specific regulatory categories for classifying probiotics, prebiotics, and synbiotics must be defined to reduce misunderstanding. They must not be treated as a single product and should be regulated differently. Moreover, probiotics or synbiotics that are altered via genetic modification should be regulated separately owing to their potential risks.
- A continuous vigilance system for long-term clinical trials of probiotics and synbiotics should exist.
- A system for reporting ongoing/encountered adverse effects of probiotic and synbiotic-based products should be practiced.
- Categorization of probiotics should also be done using an alternative framework that integrates a unique non-drug,

non-food category for some strains having probiotic potential along with two broad categories named 'probiotics as food' and 'probiotics as pharmaceuticals.' This kind of new category might make sense, which could be useful in future. Beneficial microorganisms with probiotic potential must be categorized broadly as per their intended usage and safety parameters under the following categories:

1. *Nutribiotics*, i.e., probiotics as food or beverage with specific nutritive claims only.
2. *Pharmabiotics*, i.e., probiotics as pharmaceuticals with specific health claims.

10.4 Multiple Choice Questions

1. Synbiotics are present in the marketplace as _____.
   a. food
   b. beverages
   c. pills and capsules (pharmaceuticals)
   d. All of these
2. Synbiotic-based food products in dairy matrix provide _____.
   a. healthy food products with beneficial microorganisms
   b. compromised gut health
   c. cancer prevention
   d. None of these
3. Synbiotic products in the non-dairy matrix _____.
   a. cannot be used by dairy consumers
   b. serves as an option for vegan or lactose-intolerant consumers
   c. uses matrix like ice creams, yogurts, and milk
   d. does not have any impact on the gut health of the host
4. Physiological challenges faced by synbiotics products include _____.
   a. establishment of clinical efficacy
   b. establishment of minimum effective dose
   c. Both a and b
   d. None of the above

5. Ecological challenges faced by synbiotics products include _____.
   a. compatibility of probiotics with the prebiotic components in breaking and consuming them
   b. ability of the combined system to survive in the GI tract
   c. efficiency of prebiotics as a selective substrate to probiotic microorganisms
   d. All of these

6. Processing challenges faced by synbiotics products include _____.
   a. selection of appropriate probiotic microbes
   b. different process-based and food matrix-based factors influence the development of synbiotic products
   c. maintaining sensory as well as nutritional characteristics of the final product
   d. All of these

7. Which of the following is NOT true for driving categorization pro/synbiotic products?
   a. Health claims and labeling are too specific and lack scientific evidence.
   b. False claims and gimmicks are used by companies for the promotion of their product.
   c. Multiple definitions, misleading labels, and lack of universal regulations globally.
   d. Different countries tend to follow the same rules with respect to the use of live microorganisms in any food item.

8. Which of the following statements is INCORRECT with respect to the recommendation for categorization of pro/synbiotic products?
   a. Adoption of universally accepted standard definitions and rules.
   b. Probiotics, prebiotics, and synbiotics must not be treated as a single product and should be regulated differently.
   c. Probiotics or synbiotics that are altered via genetic modification should be regulated like regular pro/synbiotics.
   d. A continuous vigilance system for long-term clinical trials of probiotics and synbiotics should exist.

9. Nutribiotics are _____.
   a. probiotics as food or beverage with specific nutritive claims only
   b. probiotics as pharmaceuticals with specific health claims only
   c. probiotics as pharmaceuticals with specific nutritive claims only
   d. probiotics as food or beverage with specific health claims only

10. Pharmabiotics are _____.
    a. probiotics as food or beverage with specific health claims
    b. probiotics as pharmaceuticals with specific nutritive claims
    c. probiotics as pharmaceuticals with specific health claims
    d. probiotics as food or beverage with specific nutritive claims

10.5 Short Answer Type Questions

Q1. Write a short note on 'commercialization of synbiotic products.'
Q2. Enlist the various challenges faced by synbiotic products.
Q3. Why does the pro/synbiotic industry need proper categorization?

10.6 Descriptive Questions

Q1. What are the challenges faced by synbiotics at the industrial level? Elaborate any one in detail.
Q2. Comment on the future recommendations required for the categorization of synbiotics.
Q3. Elaborate on the problems faced by the probiotic/synbiotic industry in terms of categorization.

10.7 Answers for MCQs

Q1	Q2	Q3	Q4	Q5	Q6	Q7	Q8	Q9	Q10
d	a	b	c	d	d	d	c	a	c

References

Arora, M. K., & Baldi, A. (2015). Regulatory categories of probiotics across the globe: A review representing existing and recommended categorization. *Indian Journal of Medical Microbiology*, *33*, S2–S10. 10.4103/0255-0857.150868

Koirala, S., & Anal, A. K. (2021). Probiotics-based foods and beverages as future foods and their overall safety and regulatory claims. *Future Foods*, *3*, 100013. 10.1016/j.fufo.2021.100013

Quintero, D., Kok, C. R., & Hutkins, R. W. (2022). The future of synbiotics: Rational formulation and design. *Frontiers in Microbiology*, *13*. 10.3389/fmicb.2022.919725

Suggested Readings

Kumar, J., Verma, S., Mazahir, F., & Yadav, A. K. (2023). Regulatory issues of synbiotics in cancer. In *Synbiotics for the Management of Cancer* (pp. 269–287). 10.1007/978-981-19-7550-9_13

Pineiro, M., & Stanton, C. (2007). Probiotic bacteria: Legislative framework—requirements to evidence basis. *Journal of Nutrition*, *137*(3), 850S–853S. 10.1093/jn/137.3.850s

11

SYNBIOTIC FOOD PRODUCTS

This chapter provides a detailed table of the steps of formulation of different dairy and non-dairy synbiotic products, along with the probiotic and prebiotic components used in their manufacture. This will allow the reader to get a crux of the processing associated with synbiotic food products (Table 11.1).

DOI: 10.1201/9781003304104-11

Table 11.1 A Summary of Synbiotic Food Products from Different Food Groups and Their Process Steps in Brief

SYNBIOTIC PRODUCT	PROBIOTICS	PREBIOTICS	STEPS OF FORMULATION	REFERENCES
DAIRY				
1. Synbiotic milk	*Lactobacillus plantarum, Lacticaseibacillus casei, Bifidobacterium animalis* subsp. *lactis* Bb-12, *Lactococcus lactis* ssp. *lactis*.	Inulin, short-chain fructooligosaccharides (sc-FOS)	The procured raw milk (cow/goat) is preheated to 55–60 °C followed by standardization. The milk is then added with prebiotics, such as inulin, in the desired quantity. After this, the mixture is heated to 95 °C for 5 min, followed by cooling (22–25 °C). Then the probiotic inoculation is done, followed by incubation (22 °C for 16–18 h), cooling, and storage.	Buran et al. (2022)
2. Synbiotic yogurt	*Streptococcus thermophilus* ATCC19258, *Lactobacillus delbrueckii* ssp. *bulgaricus* CFR2028, *Lactobacillus plantarum* CFR2194, *Saccharomyces boulardii*	Fructooligosaccharides (FOS), inulin	Fresh, pasteurized milk (3%) is preheated at 63 °C for 30 min and added with the prebiotic followed by cooling to 40 °C. The milk is then inoculated with the probiotic yogurt starter culture and kept for incubation at 40 °C for 6–8 h. After incubation, yogurt samples are stored at 4 °C.	Madhu et al. (2012); Sarwar et al. (2019)
3. Synbiotic cheese	*Lactobacillus paracasei* INIA P272, *Lactobacillus rhamnosus* INIA P344, *Lactococcus lactis* spp. *cremoris*, *Lactococcus lactis* spp. *lactis*, and *Lactobacillus acidophilus* LA5	Inulin, FOS, and galactooligosaccharides (GOS)	Pasteurized whole milk at 33 °C was added with 0.02% CaCl$_2$ in a vat container followed by the addition of the starter culture and prebiotics. After 20 min of inoculation, 0.025 g/L of Rennet was added. After 45 min, the obtained curd is cut into 2 cm cubes and heated at 37 °C for	Kavas et al. (2021); Langa et al. (2019)

(Continued)

Table 11.1 (Continued) A Summary of Synbiotic Food Products from Different Food Groups and Their Process Steps in Brief

SYNBIOTIC PRODUCT	PROBIOTICS	PREBIOTICS	STEPS OF FORMULATION	REFERENCES
			20 min. Then the whey is drained off for 30 min and the curd is distributed into cylindrical moulds. The obtained cheese is then pressed at room temperature overnight, salted in 150 g/L NaCl solution for 20 min, and at last vacuum packaged and stored to ripen at 4 °C.	
4. Synbiotic ice cream	L. delbrueckii ssp. bulgaricus, S. thermophilus, L. plantarum and L. casei, S. boulardii CNCM I-745	Inulin, FOS	The raw milk is first added to 7% sucrose and subjected to pasteurization followed by cooling and probiotic inoculation. The mixture is then subjected to fermentation for 6 h after which all the desirable ingredients, namely, prebiotics, cream, non-fat milk powder, monoglyceride, guar gum, and carboxymethyl cellulose sodium, are added and mixed (homogenized). Then the mixture is subjected to aging at 4–10 °C for 2–4 h followed by freezing at −6 °C for 10–15 min and hardening at −25 °C for 2 h. The obtained product is then kept in cold storage (−18 °C).	Elkot et al. (2022); Sabet-Sarvestani et al. (2021); Sarwar et al. (2021)
5. Synbiotic butter/ spread	S. thermophilus, B. animalis BB-12, and L. acidophilus La-5	Inulin	Inulin, emulsifier (whey protein concentrate or lecithin), and anhydrous milk fat were mixed and homogenized. Then, this emulsion is poured into container and subjected to cooling till 45 °C, followed by probiotic inoculation. After homogenization for 30 s at a rotational speed of	Toczek et al. (2022)

(Continued)

Table 11.1 (Continued) A Summary of Synbiotic Food Products from Different Food Groups and Their Process Steps in Brief

SYNBIOTIC PRODUCT	PROBIOTICS	PREBIOTICS	STEPS OF FORMULATION	REFERENCES
			$1,000$ min^{-1} the emulsions were poured into cylindrical plastic containers. The containers were kept at 42 °C for 12 h and then cooled and stored at 5 °C.	
6. Synbiotic milk powder	L. plantarum, L. casei, L. acidophilus	Inulin, xylo-oligosaccharide (XOS), FOS, and isomalto–oligosaccharide (IMO)	Two ways: (A) The prebiotic powders can be directly mixed with milk powder and subjected to packaging and storage. (B) The synbiotic milk can be dried at appropriate drying temperatures using methods like spray drying and freeze drying.	Jia et al. (2023)
FRUITS AND VEGETABLES				
1. Synbiotic juice	L. plantarum, L. casei, L. acidophilus	Fruit-specific prebiotics (FOS, GOS, etc.)	First, the juice is extracted, filtered, and prepared by diluting the fruit pulp with potable water, followed by homogenization. Then the juice is inoculated with probiotics, followed by incubation at a specific time and temperature combination based on the pH of the fruit juice. The fermented juice is then bottled and kept at refrigerated storage conditions.	Dahal et al. (2020); Mantzourani et al. (2019); Silva et al. (2021)
2. Synbiotic instant beverage powders	L. plantarum, L. casei, L. acidophilus	Fruit specific prebiotics (FOS, GOS, etc.)	Same as synbiotic milk powder	Kalita et al. (2018)

(Continued)

Table 11.1 (Continued) A Summary of Synbiotic Food Products from Different Food Groups and Their Process Steps in Brief

SYNBIOTIC PRODUCT	PROBIOTICS	PREBIOTICS	STEPS OF FORMULATION	REFERENCES
CEREALS AND LEGUMES				
1. Synbiotic milk/ beverage	*Lactobacillus brevis* PML1, *Lactobacillus casei, Lactobacillus plantarum, Bifidobacterium* spp.	Cereal/legume-specific prebiotics (raffinose-family oligosaccharides [RFOs], FOS, GOS, etc.)	Based on the type of cereals and legumes, pretreatment such as soaking or germination can be done to the cereals and legumes. Followed by this, the cereal/legume is subjected to either wet or dry grinding to extract the prebiotics in the beverage. The obtained beverage is then added with desirable flavoring ingredients and subjected to pasteurization. Then the pasteurized beverage is inoculated with probiotics in adequate quantity. The beverage is then incubated based on the type of microbial strain and pH of the beverage. The beverage is then bottled under sterile conditions and kept in refrigeration.	Chaturvedi and Chakraborty (2022b); Salmerón (2017)
2. Synbiotic yogurt	*L. delbrueckii* ssp. *bulgaricus, S. thermophilus, L. brevis* PML1	Cereal/legume-specific prebiotics (RFOs, FOS, GOS, etc.)	The cereal-based beverage is mixed with pasteurized milk or skim milk powder, pasteurized at 80–90 °C for 5 min, and cooled to 41–42 °C. Then the yogurt cultures are added, and the sample is kept in an incubator for 4–6 h. Once the desirable texture is obtained, the yogurt is stored at 4 °C.	Anino et al. (2019)
3. Synbiotic instant beverage powders	*Lactobacillus brevis* PML1, *Lactobacillus casei, Lactobacillus plantarum, Bifidobacterium* spp.	Cereal/legume-specific prebiotics (RFOs, FOS, GOS, etc.)	Same as synbiotic milk/juice powder	Chaturvedi and Chakraborty, (2022a)

References

Anino, C., Onyango, A. N., Imathiu, S., Maina, J., & Onyangore, F. (2019). Chemical composition of the seed and 'milk' of three common bean (*Phaseolus vulgaris* L) varieties. *Journal of Food Measurement and Characterization*. 10.1007/s11694-019-00039-1

Buran, I., Akal, H. C., Ozturkoğlu-Budak, S., & Yetisemiyen, A. (2022). Effect of milk kind on the physicochemical and sensorial properties of synbiotic kefirs containing *Lactobacillus acidophilus* LA-5 and *Bifidobacterium bifidum* BB-11 accompanied with inulin. *Food Science and Technology (Brazil)*, *42*, 1–8. 10.1590/fst.08421

Chaturvedi, S., & Chakraborty, S. (2022a). Comparative analysis of spray-drying microencapsulation of *Lacticaseibacillus casei* in synbiotic legume-based beverages. *Food Bioscience*, *50*(PB), 102139. 10.1016/j.fbio.2022.102139

Chaturvedi, S., & Chakraborty, S. (2022b). Optimization of extraction process for legume-based synbiotic beverages, followed by their characterization and impact on antinutrients. *International Journal of Gastronomy and Food Science*, *28*, 100506. 10.1016/J.IJGFS.2022.100506

Dahal, S., Ojha, P., & Karki, T. B. (2020). Functional quality evaluation and shelf life study of synbiotic yacon juice. *Food Science and Nutrition*, *8*(3), 1546–1553. 10.1002/fsn3.1440

Elkot, W. F., Ateteallah, A. H., Al-Moalem, M. H., Shahein, M. R., Alblihed, M. A., Abdo, W., & Elmahallawy, E. K. (2022). Functional, physicochemical, rheological, microbiological, and organoleptic properties of synbiotic ice cream produced from camel milk using black rice powder and *Lactobacillus acidophilus* LA-5. *Fermentation, 8*(4), 187. 10.3390/FERMENTATION8040187

Jia, M., Luo, J., Gao, B., Huangfu, Y., Bao, Y., Li, D., & Jiang, S. (2023). Preparation of synbiotic milk powder and its effect on calcium absorption and the bone microstructure in calcium deficient mice. *Food and Function*, *14*(7), 3092–3106. 10.1039/d2fo04092a

Kalita, D., Saikia, S., Gautam, G., Mukhopadhyay, R., & Mahanta, C. L. (2018). Characteristics of synbiotic spray dried powder of litchi juice with *Lactobacillus plantarum* and different carrier materials. *LWT – Food Science and Technology*, *87*, 351–360. 10.1016/j.lwt.2017.08.092

Kavas, N., Kavas, G., Kınık, Ö., Ateş, M., Şatır, G., & Kaplan, M. (2021). The effect of using microencapsulated pro and prebiotics on the aromatic compounds and sensorial properties of synbiotic goat cheese: Aromatic compounds and sensorial properties of synbiotic goat cheese. *Food Bioscience*, *43*, 101233. 10.1016/j.fbio.2021.101233

Langa, S., van den Bulck, E., Peirotén, A., Gaya, P., Schols, H. A., & Arqués, J. L. (2019). Application of lactobacilli and prebiotic oligosaccharides for the development of a synbiotic semi-hard cheese. *LWT*, *114*, 108361. 10.1016/j.lwt.2019.108361

Madhu, A. N., Amrutha, N., & Prapulla, S. G. (2012). Characterization and antioxidant property of probiotic and synbiotic yogurts. *Probiotics and Antimicrobial Proteins*, *4*(2). 10.1007/s12602-012-9099-6

Mantzourani, I., Terpou, A., Alexopoulos, A., Bezirtzoglou, E., Bekatorou, A., & Plessas, S. (2019). Production of a potentially synbiotic fermented Cornelian cherry (*Cornus mas* L.) beverage using *Lactobacillus paracasei* K5 immobilized on wheat bran. *Biocatalysis and Agricultural Biotechnology*, *17*, 347–351. 10.1016/j.bcab.2018.12.021

Sabet-Sarvestani, N., Eskandari, M. H., Hosseini, S. M. H., Niakousari, M., Hashemi Gahruie, H., & Khalesi, M. (2021). Production of synbiotic ice cream using *Lactobacillus casei/Lactobacillus plantarum* and fructoo-ligosaccharides. *Journal of Food Processing and Preservation*, *45*(5). 10.1111/jfpp.15423

Salmerón, I. (2017). Fermented cereal beverages: From probiotic, prebiotic and synbiotic towards Nanoscience designed healthy drinks. In *Letters in Applied Microbiology*. 10.1111/lam.12740

Sarwar, A., Aziz, T., Al-Dalali, S., Zhang, J., Din, J., Chen, C., Cao, Y., Fatima, H., & Yang, Z. (2021). Characterization of synbiotic ice cream made with probiotic yeast *Saccharomyces boulardii* CNCM I-745 in combination with inulin. *LWT*, *141*. 10.1016/j.lwt.2021.110910

Sarwar, A., Aziz, T., Al-Dalali, S., Zhao, X., Zhang, J., Ud Din, J., Chen, C., Cao, Y., & Yang, Z. (2019). Physicochemical and microbiological properties of synbiotic yogurt made with probiotic yeast saccharomyces boulardii in combination with inulin. *Foods*, *8*(10). 10.3390/foods81 00468

Silva, J. V. de C., da Silva, A. D., Klososki, S. J., Barão, C. E., & Pimentel, T. C. (2021). Potentially synbiotic grape juice: What is the impact of the addition of lacticaseibacillus casei and prebiotic components? *Biointerface Research in Applied Chemistry*, *11*(3), 10703–10715. 10.33263/BRIAC113.1070310715

Toczek, K., Glibowski, P., Kordowska-Wiater, M., & Iłowiecka, K. (2022). Rheological and textural properties of emulsion spreads based on milk fat and inulin with the addition of probiotic bacteria. *International Dairy Journal*, *124*. 10.1016/j.idairyj.2021.105217

Index